Ethnobotany of the Gitksan Indians of British Columbia

Harlan I. Smith

Edited, annotated, and expanded by

Brian D. Compton
Department of Botany
University of British Columbia
Vancouver, British Columbia, Canada

Bruce Rigsby
Department of Anthropology and Sociology
University of Queensland
St. Lucia, Queensland, Australia

Marie-Lucie Tarpent
Department of Modern Languages
Mount Saint Vincent University
Halifax, Nova Scotia, Canada

Mercury Series
Canadian Ethnology Service
Paper 132

Published by
Canadian Museum of Civilization

© Canadian Museum of Civilization 1997

CANADIAN CATALOGUING IN PUBLICATION DATA

Smith, Harlan I., 1872-1940
Ethnobotany of the Gitksan Indians of British Columbia
"Edited version of an unpublished manuscript prepared by Harlan I. Smith during the years from 1925 to 1927" — Preface
(Mercury series, ISSN 0316-1854)
(Paper/Canadian Ethnology Services, ISSN 0316-1862; no. 132)
Includes an abstract in French.
Includes bibliographical references.
ISBN 0-660-15968-6

1. Gitksan Indians — British Columbia — Ethnobotany.
2. Indians of Noth America — British Columbia — Ethnobotany.
I. Compton, Brian Douglas, 1958- .
II. Rigsby, Bruce.
III. Tarpent, Marie-Lucie, 1941- .
IV. Canadian Museum of Civilization.
V. Canadian Ethnology Service.
VI. Series.
VII. Series: Paper (Canadian Ethnology Service); no. 132.

E98.B7S64 1997   581.6'089'097011   C97-980024-2

 PRINTED IN CANADA

Published by
Canadian Museum of Civilization
100 Laurier Street
P.O. Box 3100, Station B
Hull, Quebec
J8X 4H2

Senior production officer: Deborah Brownrigg
Cover design: Roger Langlois Design

Front cover:

Clockwise:

Cornus stolonifera (Red-osier Dogwood; Figure 27)
Sedum lanceolatum (Lance-leaved Stonecrop; Figure 28)
Sorbus sitchensis (Sitka Mountain-ash; Figure 45)
Harlan Smith photographing totem pole No. 11 at Kitwanga, British Columbia, 1925. (CMC 65450)
Ledum groenlandicum (Labrador Tea; Figure 31)

Background photograph:

Athyrium filix-femina (Lady Fern; Figure 3)
Photographs (except for CMC 65450) by Jim Pojar

Back cover:

Harlan Smith transporting plaster of paris casts of petroglyphs in paper pack carrier (CMC 55792)

Page viii:

Photograph by Harlan Smith, 1925 (CMC 64376)

OBJECT OF THE MERCURY SERIES

The Mercury Series is designed to permit the rapid dissemination of information pertaining to the disciplines in which the Canadian Museum of Civilization is active. Considered an important reference by the scientific community, the Mercury Series comprises over three hundred specialized publications on Canada's history and prehistory.

Because of its specialized audience, the series consists largely of monographs published in the language of the author.

In the interest of making information available quickly, normal production procedures have been abbreviated. As a result, grammatical and typographical errors may occur. Your indulgence is requested.

Titles in the Mercury Series can be obtained by calling in your order to 1-800-555-5621, or by writing to:

> Mail Order Services
> Canadian Museum of Civilization
> 100 Laurier Street
> P.O. Box 3100, Station B
> Hull, Quebec
> J8X 4H2

Fax: 1-819-776-8300

BUT DE LA COLLECTION

La collection Mercure vise à diffuser rapidement le résultat de travaux dans les disciplines qui relèvent des sphères d'activités du Musée canadien des civilisations. Considérée comme un apport important dans la communauté scientifique, la collection Mercure présente plus de trois cents publications spécialisées portant sur l'héritage canadien préhistorique et historique.

Comme la collection s'adresse à un public spécialisé, celle-ci est constituée essentiellement de monographies publiées dans la langue des auteurs.

Pour assurer la prompte distribution des exemplaires imprimés, les étapes de l'édition ont été abrégées. En conséquence, certaines coquilles ou fautes de grammaire peuvent subsister : c'est pourquoi nous réclamons votre indulgence.

Vous pouvez vous procurer la liste des titres parus dans la collection Mercure en appelant au 1 800 555-5621, ou en écrivant au :

> Service des commandes postales
> Musée canadien des civilisations
> 100, rue Laurier
> C.P. 3100, succursale B
> Hull (Québec)
> J8X 4H2

Télécopieur : 1 819 776-8300

Abstract

This document represents an edited version of a manuscript from the Canadian Museum of Civilization on Gitksan ethnobotany prepared between 1925 and 1927 by Harlan I. Smith. It contains information on 112 botanical species and their traditional cultural roles among the Gitksan.

Résumé

Ce document est la version révisée d'un manuscrit qui appartient au Musée canadien des civilisations et qui a été rédigé par Harlan I. Smith entre 1925 et 1927. Il porte sur l'ethnobotanique des Gitksans; il contient des renseignements sur 112 espèces de plantes et sur leur rôle culturel traditionnel chez ce peuple.

ATTENTION—IMPORTANT CAUTIONARY NOTE

Some of the herbal medicines described in this report are extremely toxic, capable of causing serious illness or death if used improperly. Others, while not necessarily toxic, may cause allergenic or otherwise adverse reactions in some individuals if contacted or taken internally The information contained in this report is presented as a record of one important aspect of the cultural heritage of the Gitksan people. This document is not intended to serve as a guide for self-medication or amateur experimentation. Readers should consult an expert prior to contact with any of the plants described herein. The most toxic, or otherwise potentially injurious, plants discussed in this report are noted as such to alert readers to their hazardous nature.

Preface

The present work is an edited version of an unpublished manuscript prepared by Harlan I. Smith during the years from 1925 to 1927 while Smith was an employee of the National Museum of Canada in Ottawa, Ontario. Smith's original document has remained unpublished for nearly 70 years and is now among the collections held by the Canadian Museum of Civilization in Hull, Québec.

This edited version had its genesis during the late 1980s when I sought to consult Smith's original manuscript during a review of materials relevant to my research on Haisla (Kitamaat), Henaaksiala (Kitlope and Kemano) and Southern Tsimshian (Kitasoo) ethnobotany as part of my doctoral research project conducted in the Department of Botany at The University of British Columbia, Vancouver.

After receiving a copy of Smith's "Ethno-botany of the Gitksan Indians of British Columbia" (Smith 1925-1927), I became interested not only in what the document had to offer me in terms of relevant comparative information, but also in what the document itself represented: the earliest, relatively comprehensive ethnobotanical work for any Tsimshianic group. In fact, it is also apparently the first record of a clearly ethnobotanical investigation conducted within British Columbia as well as the first to include the term "Ethno-botany" in its title.[1] It is not, however, Smith's earliest or only manuscript focusing on the ethnobotanical features of a First Nations group of British Columbia. His efforts also include comparable works dealing with the Nuxalk (Bella Coola) and Carrier (Smith 1920-1923a, b, c, d, e, f, g).

Unfortunately, like many of Smith's related materials on file at the Canadian Museum of Civilization, his Gitksan ethnobotany document contained much redundant information and many duplicated pages. Furthermore, many of the pages contained handwritten annotations by Smith—several of which were illegible—and typed and handwritten text that had been obscured by being crossed out by Smith.[2] These features made the original manuscript very difficult to use for reference purposes.

---

[1] Another notable, yet later, ethnobotanical work was produced by Steedman (1930) who compiled Nlaka'pamux (Thompson) ethnobotanical data from field notes produced by ethnographer James A. Teit.

[2] Some clues to the nature and content of these pages come from Smith's correspondence. In a letter to Edward Sapir, Smith (1913) mentioned "trouble with the pains in my hands and legs"—an apparent reference to arthritis which may help explain why much of Smith's handwriting is so difficult to read. Smith also worked in the field from carbon copies of his notes or manuscripts, annotating them with extensive handwritten comments (cf. Smith 1922).

To overcome this difficulty I re-typed the original manuscript as a computer document to facilitate my retrieval of the information that Smith had documented. As a logical extension of this project and in recognition of several problematic aspects of the original manuscript, I decided to produce a revised version that might prove of interest and value to others. In order to increase the botanical accuracy I have provided corrected or contemporary botanical Latin species names.[3] To increase the linguistic accuracy and usefulness of the Gitksan terms recorded by Smith, I requested that Dr. Bruce Rigsby and Dr. Marie-Lucie Tarpent, who have expertise with the Tsimshianic languages, revise the transcriptions of those terms and present them in a contemporary, standardized orthography.

The result of this work is the following thoroughly revised version of Smith's manuscript on Gitksan ethnobotany with additional introductory text, ethnobotanical and linguistic commentary, and concluding remarks not present in the original manuscript. I hope that the improved ease with which Smith's manuscript may now be accessed will encourage its use and appreciation among members of the Gitksan community, interested scholars and others with an appreciation for the unique relationships between the Gitksan people and their botanical environment.

Brian D. Compton

---

[3] In some cases I have also provided ancillary comments regarding the comparable ethnobotanical features of botanical species among neighbouring First Nations groups.

Contents

List of Figures	1
Acknowledgements	3
Introduction	4
The Gitksan People, Their Language, and Their Homeland	4
Botanical Environment of the Gitksan	9
Brief Account of Harlan I. Smith's Activities at the National Museum Leading to the Production of "Ethno-botany of the Gitksan Indians of British Columbia"	11
Smith's Original Manuscript	11
Part 1. H.I. Smith's "Ethno-botany of the Gitksan Indians of British Columbia"	11
Part 2. The Gitksan Consultants and Smith's Research Methodology	13
Additional H.I. Smith Materials on Gitksan Ethnobotany and the Kitwanga Garden of Native Plants	14
Transcription of the Gitksan Ethnobotanical Lexicon	15
Botanical Species Collected as Voucher Specimens or Observed and Noted by H.I. Smith	16
Plants and Fungi Known or Used by the Gitksan	18
Fungi (Mushrooms and Their Relatives)	18
Lichens (Lichenized Fungi)	20
Bryophytes (Mosses and Their Relatives)	22
Pteridophytes (Ferns and Their Relatives)	25
Gymnosperms (Conifers and the Taxad, Western Yew)	31
Angiosperms (Flowering Plants), Dicotyledons	47
Angiosperms (Flowering Plants), Monocotyledons	138
Unidentified Species	149
Discussion, Summary, and Conclusions	149
Table 1. Summary of botanical species used as food or in food-related applications among the Gitksan	150
Table 2. Summary of botanical species regarded as animal food among the Gitksan	152
Table 3. Summary of botanical species used as material for technological and other applications among the Gitksan	153
Table 4. Summary of botanical species used as medicinal applications among the Gitksan	155
Table 5. Summary of botanical species with ritual or spiritual roles among the Gitksan	157
Table 6. Summary of botanical species with mythological roles among the Gitksan	158
Table 7. Summary of botanical species with miscellaneous cultural roles among the Gitksan	159
Table 8. Summary of botanical species recognized and named among the Gitksan but which lack cultural roles	161
References	166
Appendix 1. Writing the Gitksan Language	170
Appendix 2. Botanical species collected or observed by H.I. Smith in 1925 and 1926	171
Appendix 3. Taxa reported by H.I. Smith in individual species accounts for which no observations or collections were reported	188
Appendix 4. Botanical species that lack Gitksan names or uses, but which H.I. Smith included in the main body of his original manuscript	190
Index	203

List of Figures

Figure 1. Gitksan Territory and Settlements..8
Figure 2. Sphagnum capillaceum (Common Red Sphagnum)..........................24
Figure 3. Athyrium filix-femina (Lady Fern)...26
Figure 4. Equisetum arvense (Common Horsetail)..28
Figure 5. Lycopodium complanatum (Ground-cedar).......................................30
Figure 6. Juniperus communis (Common Juniper)..33
Figure 7. Thuja plicata (Western Redcedar)...35
Figure 8. Abies amabilis (Amabilis Fir)..37
Figure 9. Abies lasiocarpa (Subalpine Fir)...38
Figure 10. Picea x lutzii (Hybrid Sitka Spruce)...40
Figure 11. Pinus contorta (Lodgepole Pine)...42
Figure 12. Tsuga heterophylla (Western Hemlock)...44
Figure 13. Taxus brevifolia (Western Yew)..46
Figure 14. Acer glabrum (Douglas Maple)...48
Figure 15. Heracleum lanatum (Cow-parsnip)..52
Figure 16. Apocynum androsaemifolium (Spreading Dogbane).........................54
Figure 17. Oplopanax horridus (Devil's club)...56
Figure 18. Achillea millefolium (Yarrow)...58
Figure 19. Alnus rubra (Red Alder)...62
Figure 20. Betula papyrifera (Paper Birch)..65
Figure 21. Corylus cornuta (Beaked Hazelnut)...67
Figure 22. Lonicera involucrata (Black Twinberry)...70
Figure 23. Sambucus racemosa (Red Elderberry)..72
Figure 24. Symphoricarpos albus (Common Snowberry).................................74
Figure 25. Viburnum edule (Highbush Cranberry)..76
Figure 26. Pachistima myrsinites (Falsebox)...78
Figure 27. Cornus stolonifera (Red-osier Dogwood).......................................81
Figure 28. Sedum lanceolatum (Lance-leaved Stonecrop)...............................83
Figure 29. Shepherdia canadensis (Soapberry)..86
Figure 30. Arctostaphylos uva-ursi (Kinnikinnick)..88
Figure 31. Ledum groenlandicum (Labrador Tea)...90
Figure 32. Oxycoccus oxycoccus (Bog Cranberry)...92
Figure 33. Vaccinium alaskaense Howell (Alaskan Blueberry).........................94
Figure 34. Vaccinium ovalifolium (Oval-leaved Blueberry)..............................95
Figure 35. Vaccinium caespitosum (Dwarf Blueberry)....................................97
Figure 36. Lupinus arcticus (Arctic Lupine)..100
Figure 37. Nuphar polysepalum (Yellow Pond-lily).......................................105
Figure 38. Amelanchier alnifolia (Saskatoon)...111
Figure 39. Fragaria virginiana (Wild Strawberry)...114
Figure 40. Malus fusca (Pacific Crab Apple)..116
Figure 41. Prunus pensylvanica (Pin Cherry)...118
Figure 42. Prunus virginiana (Choke Cherry)...120
Figure 43. Rubus idaeus (Red Raspberry)...123
Figure 44. Rubus spectabilis (Salmonberry)..126
Figure 45. Sorbus sitchensis (Sitka Mountain-ash).......................................128
Figure 46. Populus balsamifera (Black Cottonwood)....................................131
Figure 47. Castilleja miniata (Common Red Paintbrush)...............................135

Figure 48. <u>Urtica</u> <u>doica</u> (Stinging Nettle)..137
Figure 49. <u>Calla</u> <u>palustris</u> (Wild Calla)...139
Figure 50. <u>Fritillaria</u> <u>camschatcensis</u> (Northern Rice-root)..................................143
Figure 51. <u>Veratrum</u> <u>viride</u> (Indian Hellebore)..147

Acknowledgements

We wish to acknowledge and thank the following individuals for their valuable contributions to the production of this document: Gloria Bishop, Skooker Broome, Nicole Chamberland, Dennis Fletcher, Manon Guilbert, Chris Kirby, Andrea Laforet, M.J. Shchepanek, Barbara Sterritt, Benoît Thériault, and Margery Toner. In addition, we wish to thank the members of the Sim'algax Working Group, Leslie M. Johnson Gottesfeld and Jim Pojar, who reviewed the draft version of this document and provided valuable editorial commentary. The plant photographs that appear in the following pages were taken by Jim Pojar who kindly provided them for use in this publication.

Introduction

Harlan I. Smith did not present any introductory discussion to his Gitksan manuscript with the exception of a single sentence regarding medicinal plants: "Plant remedies were not family secrets among the Gitksan Indians, according to Luke Fowler, May 24, 1926, (as among the Bella Coola H.I.S.),[4] but all the people knew of all the remedies." His interest in herbal medicines prompted his production of a published paper (Smith 1929) on the subject of Nuxalk, Carrier, Sekani and Gitksan materia medica, with emphasis on botanical agents. According to Smith's (1929:47) comments in that paper:

> The method the writer employed was to submit specimens of the plants to his informants and question them concerning all their uses. Of the information thus obtained only the medicinal portion has been presented here. The plants were later identified by Mr. M.O. Malte, of the National Museum, Ottawa.

Smith's research on Gitksan ethnobotany resulted in his production of 250 pages involving the documentation and discussion of Gitksan names, uses and other cultural roles for over 100 plants and fungi. Smith mentioned several additional botanical taxa in his original manuscript although they did not have Gitksan names or ethnographic data associated with them. The results of Smith's work are presented in the following pages together with additional descriptive or explanatory materials and, on occasion, information extracted from other written materials, including those that Smith produced.

The Gitksan People, Their Language, and Their Homeland

Smith did not provide any discussion of the Gitksan people in his manuscript on Gitksan ethnobotany. He did, however, prepare chapters on "Indians," "The Indian Culture of the Pacific Coast of Canada," and "The Gitksan" in a manuscript he prepared as a guide to totem poles that could be observed in Kitwanga and Kitseguecla by passengers on the Canadian National Railway (Smith 1926). The text that Smith prepared as Chapter 7 for that document is presented below:[5]

---

[4] Smith sometimes used his initials ("H.I.S.") to designate comments that he included parenthetically among those of the Gitksan individuals he consulted. Additional comments regarding Smith's use of "H.I.S." are presented in the section on the "Transcription of the Gitksan Ethnobotanical Lexicon."

[5] This passage represents the typed version of Smith's discussion of the Gitksan as it appears in Smith 1926. A few minor handwritten annotations appearing in the original document have not been incorporated into this reproduction of Smith's text.

THE GITKSAN

The Gitksan are one of the three dialectic divisions of the Tsimshian linguistic stock. They live in British Columbia along the upper waters of Skeena River and its tributaries, in fact Gitksan means People of smooth waters, Ksan being the name of the river.[6] Immediately to the east of them, near New Hazelton on the Canadian National Railway, are the Carriers of the great Athapascan[7] linguistic stock, totally different in language and culture. The Gitksan, are divided into four phratries which have for their crests the Raven, Fireweed, Wolf, and Eagle. In the old days several related families lived in one large house.

The Gitksan villages are Kaldo, on the Skeena River, sixty-one miles north of Hazelton, Kiskagas, on the Babine River, forty-nine miles north of Hazelton, Kispiox, at the junction of the Kispiox and Skeena Rivers, eight miles north of Hazelton, Gitanmax at Hazelton, Kitwanga, a station on the Canadian National Railway, Kitwinkul, fourteen miles north of Kitwanga, Gitsegyukla, on the south beach of the Skeena River, one mile below Skeena Crossing, also a station on the Canadian National Railway, the modern mission town of Minskinisht, on the south side of the river opposite Cedarvale station, and Glen Vowell, another mission village, four miles north of Hazelton. The total population of the Gitksan Indians in 1904 was 1120.

During the salmon canning season, in July and August, most of these Indians go to the coast, where they join the fishing fleet while the women work in the canneries. They may now be seen at nearly all the stations of the Canadian National from Hazelton west to Prince Rupert.

The brevity of Smith's comments on the Gitksan suggest the appropriateness of including here an expanded discussion of the culture, language, homeland and environment of the Gitksan. For this reason a more detailed discussion of the Gitksan is presented in the following paragraphs:

---

[6]This means, literally, 'people of Xsan (Skeena River).'
[7]This is also spelled "Athabaskan."

The Gitksan people speak an indigenous language which they call Sim'algax̱, "the real or true language." Their Nisg̱a'a neighbours, who live in the lower Nass Valley, speak a separate, but mutually intelligible language which they also call by the same name. Together the Gitksan and Nisg̱a'a languages comprise the Interior Division of the Tsimshianic language family.

The Tsimshian people, who live downstream on the Skeena and on the coast, speak a third indigenous language, which they also call Sim'algax̱, but it differs significantly from the Gitksan and Nisg̱a'a languages—Gitksan people cannot understand it until they learn it first.

In past times, the people at Hartley Bay and Klemtu (from the old village of Disjuu or China Hat) spoke a fourth indigenous language, which they call Sgüüx̱s or Sgüümx̱. Only a few speakers of Sgüüx̱s remain. It is more like the Tsimshian than the Gitksan or Nisg̱a'a languages, and together, it and Tsimshian comprise the Maritime Division of the Tsimshianic language family.

It is readily apparent even to the casual observer that the four Tsimshianic languages are closely related and belong to the same language family, but a great deal of historical linguistic research has not turned up evidence that the Tsimshianic languages are genetically related to any other language families of the Northwest Coast or neighbouring culture areas. All the evidence indicates that the Tsimshianic-speaking people are old residents of the region who have long participated in the borrowing of words and other elements of language with neighbouring peoples.

The Gitksan people have long had close cultural and social relations with their eastern neighbours, who speak a language called Babine or Witsuwit'en—it belongs to the widespread Athabaskan language family. The Gitksan have borrowed some words for animals from their Babine and other Athabaskan neighbours, while the latter have in turn borrowed some features of social organization from the Gitksan, as well as many personal names and some botanical names.

The Gitksan homeland covers an area that contains portions of four major river systems (see Figure 1). These are the Skeena River from above Terrace, the middle and upper Nass River, the Kispiox River, and the Babine River (the last two are major tributaries of the Skeena). The Skeena is the predominant watercourse of the Gitksan homeland, providing a direct route to coastal and interior groups with whom the Gitksan interacted and traded. In the past century, most of the Gitksan winter villages were situated along the Skeena River, and indeed the name Gitksan (better, Gitxsan in the eastern dialects and Gitxsen in the western ones)[8] means "people of the Skeena River."

---

[8]The dialect label Western Gitksan is abbreviated in this document as WG, while Eastern Gitksan is abbreviated as EG. Western Gitksan, sometimes referred to as the Geets' ("downriver region") dialect, is spoken at Kitseguecla, Kitwancool and Kitwanga. Eastern Gitksan, sometimes referred to as the Gigeenix ("upriver region") dialect, is spoken at Gitanmax and Kispiox, and formerly was spoken at the now uninhabited communities of Kisgegas and Kuldo. Contemporary dialectal variants of many Gitksan botanical terms are provided in this document although Smith's transcriptions give little or no indication of such dialectal variation which may represent linguistic changes that have taken place since the time of Smith's work on Gitksan ethnobotany.

Gitksan Territory and Settlements

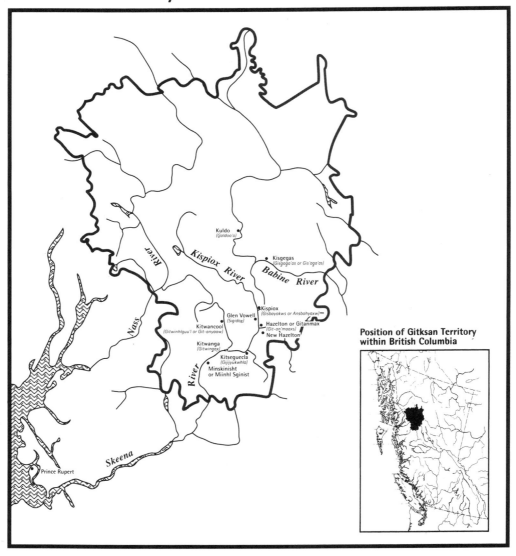

Figure 1. Gitksan Territory and Settlements

The main contemporary Gitksan communities are Kitwanga (Gitwingax), Kitseguecla (Gijigyukwhla), Kitwancool (Gitwinhlguu'l, but now its people prefer its older name, Gitanyaaw), Hazelton or Gitanmax (Git-an'maaxs), Glen Vowell (Sigidox), and Kispiox (Gisbayakws or Ansbahyaxw). Earlier this century, people still lived in Kisgegas (Gisgaga'as or Gis'aga'as) and Kuldo (Galdoo'o).

The main units in Gitksan social structure are the house groups (wilp), which are resource-owning, -managing, and -using corporations. There are perhaps fifty or so active Gitksan house groups, and they are important in social, ceremonial, economic and political life. Gitksan houses and their members also belong to one or another of four exogamous matrilineal phratries (pdeek, called "tribes" or "clans" locally). These are Fireweed/Killerwhale (Gisk'ahaast or Gisk'aast), Frog/Raven (Ganeda in the west, Laxsee'l in the east), Wolf (Laxgibuu), and Eagle (Laxsgiik).

Over the past century, the Gitksan have sought to get recognition in Canadian law of their continuing Aboriginal title to their homeland. With their Wet'suwet'en neighbours, they brought an action in the Supreme Court of British Columbia against the provincial and federal governments in the late 1980s. Following an unfavourable decision in 1991, they appealed the Delgamuukw case in the Court of Appeal and got a decision that native rights had not been extinguished to the degree found earlier. They then took the case on appeal to the Supreme Court of Canada. In early 1994, the Supreme Court gave the Gitksan and Wet'suwet'en hereditary chiefs leave from their action for two years, and they signed an accord with the British Columbia and Canadian governments to negotiate a treaty.

Botanical Environment of the Gitksan

The Gitksan village sites and areas used for hunting, fishing and the gathering of various other biotic and abiotic resources are situated within one or another of seven of the 14 biogeoclimatic zones of British Columbia. The zones that are encompassed by Gitksan territory are: Alpine Tundra, Spruce-Willow-Birch, Boreal White and Black Spruce, Sub-Boreal Pine-Spruce, Sub-Boreal Spruce, Engelmann Spruce-Subalpine Fir and Interior Cedar-Hemlock. Some additional zones located adjacent to Gitksan territory, such as the Coastal Western Hemlock zone, were familiar to many Gitksan from their travels to the coast to collect resource items or obtain them through trade with various coastal groups. These zones are relatively homogeneous in climatic terms and are further characterized by the presence of biological, especially botanical, assemblages that are distinctive for each zone. A map with more complete details of the characteristics of these zones may be found in MacKinnon et al. (1992).

The common botanical species of Gitksan territory area include a number of terrestrial and aquatic, vascular and non-vascular species. Trees are by far the most conspicuous plants in the region and conifers predominate in the forested areas. These include western redcedar, amabilis fir, subalpine fir, several species of spruce including hybrid Sitka spruce, lodgepole pine and western hemlock. Still other conifers which may not occur directly within Gitksan territory, were known to and used by the Gitksan. These include yellow cedar, Sitka spruce, and western yew. Several deciduous trees or tall shrubs also are common in or near Gitksan territory and were known to or used by the Gitksan: Douglas maple, several species of alder, paper birch, black hawthorn, Pacific crab apple, bitter cherry,[9] trembling aspen, black cottonwood, and several species of willow.

Several common shrubs are characteristic of the Gitksan territory and are known for their roles in Gitksan culture. These include devil's club, beaked hazelnut, red elderberry, common snowberry, highbush cranberry, falsebox, red-osier dogwood, soapberry, kinnikinnick, Labrador tea, various species of blueberries, bog cranberry, various species of gooseberries and currants, saskatoon, several rose species, red raspberry, thimbleberry, trailing raspberry, salmonberry, Sitka mountain-ash, and western mountain-ash.

The Gitksan territory is also home to numerous herbaceous plant species. Of these, many are common and were used for one or another purpose by the Gitksan. These include cow-parsnip, yarrow, bunchberry, yellow pond-lily, fireweed, wild strawberry and stinging nettle, skunk-cabbage, nodding onion, and northern rice-root to name a few. Other botanical species that are common to the region include numerous fungi and lichens (lichenized fungi), bryophytes (e.g., sphagnum mosses, and many others) and pteridophytes (e.g., wood fern, lady fern, licorice fern and various horsetails and clubmosses).

---

[9]This species occurs probably only in the extreme southwestern portion of Gitksan territory, if at all.

Brief Account of Harlan I. Smith's Activities at the National Museum Leading to the Production of "Ethno-botany of the Gitksan Indians of British Columbia"

As early as April 1925 members of the Board on Preservation of Totem Poles at the National Museum of Canada[10] initiated efforts to engage Smith in a project to repair, erect and preserve totem poles in the neighbourhood of Hazelton, British Columbia (Sapir 1925) "for the benefit of science, art, and tourist attraction for the C.N.R. [Canadian National Railway]" (Smith 1936). By early May 1925 Smith had been formally notified of his intended role in this project (Bolton 1925). The Annual Report for 1926 indicates that Smith was involved from May to October 1926 in work dealing with the conservation of totem poles located in proximity to the Canadian National Railway (Collins 1928a; also see Smith 1925-1926, 1926, 1928).

During the period of his totem pole work, Smith also engaged in research on the archaeology and material culture of the Gitksan and Carrier (Collins 1928a:7). These ethnographic investigations represented a continuation of work which he began in 1925 among the Gitksan and complemented Smith's research on Nuxalk and Carrier ethnobiology which he began in 1920 (Smith 1920-1923a, b, c, d, e, f, g). Smith continued his work on totem pole conservation in Hazelton, British Columbia during 1927 at which time he also made motion picture records of Gitksan, Tsimshian and Carrier ethnographic subjects (Collins 1929a:7, 8).

Smith apparently had concluded his investigation of Gitksan ethnobotany by 1926 or 1927 (see Smith 1929:47) since by 1928 he was on his way to southern British Columbia to collect archaeological and ethnological specimens and to make additional documentary films among the Ktunaxa (Kootenay) and Secwépemc (Shuswap) of British Columbia, and the Blackfoot of Alberta (Collins 1929b:3).

Smith's Original Manuscript
Part 1. H.I. Smith's "Ethno-botany of the Gitksan Indians of British Columbia"

From the consultant information provided by Smith it is evident that the majority of ethnobotanical data he recorded were derived from individuals associated with the Gitksan

---

[10]Alcock, Chief Curator of the National Museum of Canada, wrote in the forward to the 1954-55 Annual Report (Alcock 1956) that "The National Museum of Canada may be said to date from 1841 when at the first session of the first United Parliament of Upper and Lower Canada an Act was passed voting money to establish a Geological Survey." Up until 1927 the Museum was officially referred to as the Victoria Memorial Museum since it was housed in the Victoria Memorial Museum building as a separate branch of the Geological Survey of Canada, which also was housed in the same building.

community of Kitwanga. The names and uses discussed by these Gitksan consultants are undoubtedly indicative of widespread Gitksan ethnobotanical features but may also reflect more specific community-based or personal knowledge. Only additional research with contemporary Gitksan consultants can clarify this and other questions regarding Gitksan ethnobotany and add to the list of species with known Gitksan names, uses, and beliefs associated with them.

Because of this, the present document should not be considered as a complete or final record of Gitksan botanical knowledge and use. Rather, it represents an attempt to increase the legibility, accessibility and usefulness of the earliest known work on the subject. The presentation of this edited version is intended to benefit interested members of the contemporary Gitksan community and to serve as a reference for continuing Gitksan and comparative ethnobiological studies.

Several points regarding the original and edited versions of this manuscript should be mentioned here. In general, an effort has been made to retain the original character of Smith's manuscript. Many passages are reproduced here exactly as they were originally presented. In other cases, the text has been edited to conform to the style of presentation used elsewhere in the text. In some of these cases the phrase "According to Smith..." may precede the edited text. Some data have been summarized to eliminate or reduce the redundancy found in some portions of the original text, including instances where two or more Gitksan consultants are credited with identical information. More substantial editorial comments, including relevant data presented elsewhere in the original manuscript or cited from other sources (Smith 1927, 1929), appear in the main text of this edited version in (if they represent primary data) or in footnotes (if they represent ancillary data).

Throughout the text various errors and inconsistencies of spelling, punctuation, and grammar have also been corrected. In addition, there are several sections that Smith did not include in the original text. These sections include "Introduction;" "The Gitksan People, Their Language, and Their Homeland;" "Brief Account of Harlan I. Smith's Activities at the National Museum Leading to the Production of 'Gitksan Ethno-botany';" the present section dealing with "Smith's Original Manuscript (Parts 1 and 2);" "Additional H.I. Smith Materials on Gitksan Ethnobotany and the Kitwanga Garden of Native Plants;" and "Transcription of the Gitksan Ethnobotanical Lexicon."

A list of voucher materials provided by Smith is presented under the section entitled "Botanical Species Collected as Voucher Specimens or Observed and Noted by H.I. Smith." The main body of the text has been presented in the section entitled "Plants and Fungi Known or Used by the Gitksan." Finally, sections involving "Discussion, Summary and Conclusions," "References," several appendices and an index have been added to this edited version.

Part 2. The Gitksan Consultants and Smith's Research Methodology

Smith provided some information on his methods and the Gitksan consultants with whom he worked in the 1929 document, "Materia Medica of the Bella Coola and Neighbouring Tribes of British Columbia":

> The information relating to the Gitksan was secured in 1925-26 from the late John Fowler and Abraham Fowler of Kitwanga, Luke Fowler and Bob Robinson of Hazelton, with additions and corroboration from a few others, all old Gitksan Indians, apparently full-blood. The information was secured mainly in English, although Abraham Fowler used also the Chinook jargon.

In his original manuscript on Gitksan ethnobotany (Smith 1925-1927), Smith identified the following seven individuals who reportedly provided information regarding Gitksan ethnobotany:

John Fowler
    (of Kitwanga; born in Hegwilget[11] according to Luke Fowler; his father was born at Kispiox and his mother was from Kitwanga)

Abraham (Abe) Fowler
    (of Kitwanga; born in Hegwilget and over 60 years old according to Luke Fowler; his grandmother was from Kitwancool, his father was from Hegwilget, and his mother was born in Kispiox)

Luke Fowler
    (of Hazelton; 49 years old in 1926; his father and mother were from Kitwanga)

Bob Robinson (of Hazelton)

Johnson (John) Laknitz. (Laknitz)[12]

---

[11] This is also spelled "Hagwilget."
[12] Altogether, Smith gave three alternative spellings for this name in his original manuscript: Laknitz, Laknítz and Lacknitz. In the present edited version, "Laknitz" is used throughout the text.

(about 49 years old in 1926 according to Luke Fowler; his father, Jim
Laknltz,[13] and mother were from Kitwanga)

Gus Sampare and Mrs. Gus Sampare[14]

Robert A. Sampare and Mrs. R.A. Sampare

W.C. Washburn[15]

In addition to the seven individuals who Smith named for inclusion in a preface to his manuscript, Smith also mentioned the following five individuals: Frank Clark,[16] a man identified only as Joshua,[17] a man identified only as "Brown, a Gitksan Indian at Hazelton, B.C.,"[18] and two unidentified women.[19]

Additional H.I. Smith Materials on Gitksan Ethnobotany and the Kitwanga Garden of Native Plants

Another small body of information on Gitksan ethnobotany was prepared by Smith in addition to his "Ethno-botany of the Gitksan Indians of British Columbia." One item, a small handbook containing information on a small number of species, was published by the National Museum of Canada and the Department of Indian Affairs in 1927. This document is limited in size and content and contains information excerpted from Smith's Gitksan ethnobotany manuscript.[20] Some information from this handbook has been included in the present edited document.

Associated with this handbook are additional related items: (1) an undated typed cover page with an alternative title; (2) a collection of several additional handwritten pages entitled

---

[13] Elsewhere, Smith (1925) identified Jim Laknitz as "one of the two leaders in the village of Kitwanga."

[14] Although Mrs. Gus Sampare is mentioned here, no data are associated with her name in Smith's original manuscript. It is likely that she confirmed data attributed to her husband.

[15] Elsewhere, Mr. Washburn was identified as having been in charge of the Gitksan workers involved in restoration efforts focused on totem poles at Kitwanga (Anonymous 1926:15). No data are associated with his name in Smith's original manuscript.

[16] Smith further commented that "according to Gus Sampare and R.A. Sampare [Mr. Clark] is good for Ethno-Botany" but no data are associated with his name in Smith's original manuscript.

[17] This is likely Joshua Moody, a Nuxalk ethnobotanical consultant who had worked with Smith prior to 1925. No data are associated with his name in Smith's original manuscript.

[18] This individual remains to be conclusively identified. He is mentioned with reference to comments about Salix sp., willow (#93).

[19] These individuals remain to be conclusively identified. They are mentioned with reference to comments about Corydalis aurea ssp. aurea, golden corydalis (see Appendix 4, #30).

[20] This document has been photographed and included on Microfiche No. VII-C-83M (B90 F3) of the Harlan I. Smith materials available from the Canadian Museum of Civilization. The microfiche copy contains a cover page and only seven additional pages containing information on seven species of plants used by the Gitksan: yarrow, hazelnut, black twinberry, snowberry, saskatoon, black hawthorn and stinging nettle.

"Preliminary Catalogue of the Kitwanga Garden of Native Plants" (Smith n.d.); and (3) a letter from D.F. Davies, Director of the Field Museum of Natural History. The first two of these items seem to represent materials that Smith prepared prior to the production of the printed handbook. The last of these items—the D.F. Davies letter written in 1926 to Smith—deals with a request by Smith for information on the availability of identification labels applicable for native plants of British Columbia. Smith undoubtedly wanted to obtain labels for outside use to identify plants discussed in his handbook. Additional clues to the intended use of the handbook are contained in handwritten comments regarding the natural growth of some species "on the tourist path through the Outdoor Totem Pole Museum" and comments on the date of planting of other species or their occurrence "in the [Kitwanga] district." Little additional information regarding a native plant garden in Kitwanga is known[21] but the totem pole display discussed by Smith is represented by a set of two blueprints in the Smith Collection at the Canadian Museum of Civilization (Smith 1925-1926).

Transcription of the Gitksan Ethnobotanical Lexicon

Smith stated in his original manuscript that "(m)y spelling of Indian words is very problematical, because some of the sounds I can neither hear nor make, and I am unable to record adequately many of those that I can hear."[22] As languages go, Gitksan is a typical Northwest Coast language that has complex consonant sounds and clusters of consonants not found in English or other European languages. It is not an easy language for the untrained linguist or anthropologist to hear, record and analyze, so we should not expect that Smith would have done an adequate job of recording Gitksan words and forms. Nonetheless, Smith did record Gitksan names for the majority of species included in his Gitksan manuscript.

In the original manuscript, several Gitksan botanical terms (and other items) were followed by the initials, "H.I.S." which Smith apparently used to denote (1) transcriptions produced according to his personal phonetic reckoning using a non-technical orthography, (2)

---

[21] By 1926, a published reference appeared mentioning that a group of four Gitksan boys was formed as Troop I of the Skeena Boy Guards. They were charged with the tasks of providing "conspicuous service in guarding the totem poles [at Kitwanga] and the plant specimens in the Kitwanga garden of native plants" (Anonymous 1926:15).

[22] Before embarking upon his research on Nuxalk ethnobiology in 1920, Smith expected to receive linguistic assistance from Edward Sapir, then Chief of the Division of Anthropology at the Geological Survey of Canada. For various reasons Sapir was unable to accompany or meet Smith in Bella Coola, B.C., and therefore could not offer his expertise in transcribing Nuxalk plant and animal names. In a letter dated 9 July 1920, Smith (1920) wrote to Sapir that: "I am sorry to learn that you would not be here with me first because I wanted your companionship and second because you could help me greatly with my work with the Indians by taking down the names of plants and animals. I make a very poor job of it and I often wish to refer to a plant or animal when it is not present and so I need its name." Later, when he engaged in similar work among the Gitksan, Smith continued to rely upon his own ability to write Gitksan words.

English plant names that are at variance with the reference work that Smith consulted for botanical nomenclature (i.e., Henry 1915), possibly indicating locally common English names or names more familiar to Smith than those reported by Henry, or (3) comparative statements based on his other ethnobotanical investigations (e.g., among the Nuxalk). Where Smith provided alternative transcriptions of Gitksan terms with "H.I.S." following a term in parentheses, the term has been retained in parentheses but the "H.I.S." has been omitted.[23] All alternative forms of Gitksan terms originally reported by Smith, either typed or handwritten, have been reported in the present edited version following the phrase "Gitksan Name." Generally they appear in the relative order in which they appeared in the original manuscript, they are written using the orthographic representation that Smith indicated and they include all the variations that Smith provided.[24]

Bruce Rigsby and Marie-Lucie Tarpent have examined and considered Smith's renderings of these Gitksan words and wherever possible, they have retranscribed, or redacted and recast them in the modern Gitksan practical orthography (see Appendix 1) to make them accessible to students and speakers of the Gitksan language. Many of the words are known from other sources, while it is not possible to recognize others from Smith's transcriptions, and a number of species appear not to have conventional names, but the Gitksan consultants probably gave them nonce descriptions, e.g., see ishabasxwit (under entry #40). It is likely that Smith's consultants provided him with some ostensive definitions of plants in their natural habitats, e.g., as "pretend-grass" or "one that bears gather," and he wrote them down partially.[25]

Botanical Species Collected as Voucher Specimens or Observed and Noted by H.I. Smith

Smith collected a number of botanical specimens as vouchers for his ethnobotanical research among the Gitksan. In other cases, he observed, noted and possibly presented specimens to one or more Gitksan consultants for identification, but he did not retain those specimens. The extent of his collection and observation seems evident from the individual species accounts throughout the text of his original manuscript and two lists of collections or

---

[23] All other instances of Smith's use of "H.I.S." have been omitted from the present edited document.
[24] Smith did not use a standardized or consistent orthographic system in his transcription of Gitksan words. In some cases Smith used capital letters in writing Gitksan words but this capitalization was inconsistent and apparently not meaningful to the proper representation of the words. Therefore, in the plant names reproduced in the document the capitalization has not been retained.
[25] Where Gitksan words are translated, single quotation marks are used to denote literal translations and double quotation marks to indicate approximate English glosses.

observations for the years 1925 and 1926 (see Appendix 2). Smith reported 28 taxa (species or genera) as collected or observed during 1925 and 114 species during 1926. These collections or observations account for a total of 122 distinct taxa with 20 taxa (species or, in some cases, genera only) reported for both years. In the main body of the manuscript, 165 botanical taxa are listed. Of those, 22 taxa were not associated with collections or observations reported by Smith (see Appendix 3). Botanical species listed and discussed by Smith that lack Gitksan names or cultural roles are presented in Appendix 4.

Except for the lichens sent to G.K. Merrill in Maine for identification, all of Smith's collections were sent for identification to Dr. M. Oscar Malte, Chief Botanist in the Division of Biology of the National Herbarium of Canada from 1921-1933.[26] The scientific identifications associated with taxa reported by Smith as observed, but not collected, may have been proposed by Smith since there is no direct evidence from his Gitksan manuscript that a botanist was involved in these identifications.

The numbers that accompany the botanical taxa in Smith's original Gitksan manuscript often consist of one number followed by a second in parentheses. In those cases, the first number apparently represents Smith's collection number for that species (or simply a number employed in some other method of documenting the specimens), and the page of the corresponding species description in Henry (1915). For example, Smith referred to "81(13) Smilacina racemosa L., False Solomon's Seal," in which case the first number, 81, refers to page 81 in Henry's work and the second number, (13), refers to Smith's thirteenth collection or observation. In the present document there are also cases where multiple collection or observation numbers are listed together, e.g., "(114) (136) (143) Angelica genuflexa Nutt." Most of the species mentioned in the original manuscript are named according to the nomenclature presented by Henry although some species names presented by Smith do not appear in Henry's flora and may have been provided by Malte.[27]

---

[26] It seems that none of the botanical specimens that Smith may have collected for his Gitksan research survive today. The only known botanical specimens attributed to Smith comprise the Harlan I. Smith Collection of the National Herbarium of Canada at the Canadian Museum of Nature in Ottawa, Ontario. This collection consists of 225 specimens of vascular plants collected by Smith from 1920 to 1923, before he began his Gitksan research. Those specimens are identified as having been collected in relation to Smith's (1929) work on the materia medica of the Nuxalk and some of their First Nations neighbours (P. Frank, pers. comm. 1995). No additional botanical specimens attributed to Smith are known to exist among any of the various botanical collections of the National Herbarium of Canada (P. Frank, pers. comm. 1995) and none of the known Smith specimens are associated with Smith's Gitksan research (M.J. Shchepanek, pers. comm. 1995).

[27] In addition, several animals are mentioned by their English names throughout the text of Smith's original manuscript. Where appropriate, their likely zoological Latin and Gitksan identifications are provided in footnotes in the present document. The zoological Latin terminology in presented in accordance with Cannings and Harcombe (1990).

Plants and Fungi Known or Used by the Gitksan

In the present edited version of Smith's manuscript, the original scientific botanical names used by Smith have been retained with Smith's spelling errors corrected (in the case of plant names, the corrections have been made to conform to Henry 1915). In the few cases where Smith did not include a botanical binomial to accompany the common English name presented for a species, the nomenclature applied to that species by Henry has been included. In addition, contemporary botanical synonyms and common English plant names in accordance with Douglas et al. (1989, 1990, 1991, 1994) have been included for each taxon where appropriate.

Smith's data are arranged here alphabetically by species name[28] within the major categories of fungi (mushrooms and their relatives), lichens (lichenized fungi), bryophytes (mosses and their relatives), pteridophytes (ferns and their relatives), gymnosperms (conifers and the taxad, western yew), and angiosperms (flowering plants). The pteridophytes, gymnosperms and angiosperms (subdivided into dicotyledons and monocotyledons) are subdivided by families listed in alphabetical order as presented in Douglas et al. (1989, 1990, 1991, 1994) using common family names as presented by Taylor and MacBryde (1977). Several taxa previously treated as species are now subdivided into subspecies and, in some cases, varieties. In those instances, the subspecies or varieties most likely to be encountered within Gitksan territory are presented first, with the related subspecific taxa that may occur only sporadically or in restricted areas within the Gitksan territory listed last. Where more than two subspecific taxa are listed by Douglas et al. (1989, 1990, 1991, 1994) for any given species only the species name is presented here.

Fungi (Mushrooms and Their Relatives)

?Clavariaceae (Coral Fungus Family) or Hydnaceae (Tooth Fungus Family) (Order
    Aphyllophorales [Coral and Pore Fungi and Allies])

---

[28]Where species are unidentified in botanical terms, entries may appear in alphabetical order of tentatively identified genera. Or, they may appear at the end of a particular major botanical category as an unidentified species affiliated with that major botanical category.

(1) Unidentified Fungus, possibly one or more species of <u>Clavaria</u> or <u>Ramaria</u> (Coral Fungi), or <u>Hericium</u> <u>coralloides</u> (Scop.) Pers. (Coralloid Hericium) or, possibly, <u>Hydnum</u> sp. (Tooth Fungus)

Identified by Smith as: (141) a large beautiful fungus resembling a coral or sponge; large fungus-like sponge beautiful, brown on dead tree

Modern spelling of Gitksan name: lixsgedim gayda ts'uuts' (WG), lixsgadim gayda ts'uuts' (EG) ("a different mushroom," literally, 'a different hat of bird')[29]

Smith's transcription of Gitksan name: lĭlskadam kaedatūts (Lĭlska means different, kaedatūts is the name given to puffballs and mushrooms); [30] liɫskadam kae dat ūte (different = liɫshat, fungus = kae dat ūts)

Modern spelling of Gitksan name: gayda ts'uuts' (WG), gayda ts'uuts' (EG) (literally, 'hat of bird,' or 'bird-hat')

Smith's transcription of Gitksan name: kaedatsots (kydatsoats)

Among the Gitksan, according to Bob Robinson, September 5, 1926, this plant is said to grow on dead trees and to be of no use. Another Hazelton man called it kaedatsots (kydatsoats) and said it was of no use and had no story about it.

?Polyporaceae (Pore Fungus Family)

(2) ?<u>Inonotus</u> <u>obliquus</u> (Pers.: Fr.) Pilat (Cinder Conk)

Identified by Smith as: fungus on birch or hemlock

Smith's transcription of Gitksan name: dī dīyuh

Modern spelling of Gitksan name: didihuxw[31]

The fungus, dī dīyuh, that grows out of a birch (<u>Betula</u> <u>papyrifera</u>) and sometimes out of a hemlock, was lighted and used to sear a rheumatic person. This burning fungus was also applied by little boys and girls to themselves to learn if they would make good husbands and

---

[29] The Gitksan call mushrooms and toadstools by a form—gayda ts'uuts'—that literally means 'hat of bird' or 'bird-hat.' It recognizes the similarity between the umbrella shape of a prototypical mushroom and the characteristic rain hat of Northwest Coast indigenous people. The form above is probably a descriptive nonce phrase.

[30] Smith recorded the same term—gayda ts'uuts'—for several fungi, including this one. The term has cognates in Coast Tsimshian and Nisga'a that refer to mushrooms and which mean, literally, 'bird's hat' (cf. Dunn 1978). Perhaps here Mr. Robinson—the Gitksan individual who seems to have provided this term—was attempting to indicate to Smith that the fungus in question was different from other fungi (mushrooms and possibly also puffballs) that are referred to as gayda ts'uuts'.

[31] Gottesfeld (1992) has reported the terms tiiuxw and mii'hlw for this species along with Gitksan uses comparable to those reported by Smith for the "fungus on birch or hemlock." The Nisga'a use an unidentified fungus from hemlock called dihuxwt for moxibustion. Elsewhere, Smith (1920-1923g) reported that the Ulkatcho Carrier used an unidentified fungus called "dee you, or dîîu" as kindling.

wives, or would talk back. Those that stood it and did not quickly take it away were said would be good and not talk back. Those that took it away quickly would be the opposite. A group of five or six children might try this prophesying game. Luke Fowler said when he did it he left the brand on until he was burned.

Unidentified Family (?Polyporaceae)

(3) Unidentified Fungus[32]

Identified by Smith as: fungus that grows on hemlock; collected August 1, 1926 (no collection number given)

Smith's transcription of Gitksan name: kaet dă tsōōts (kaet = hat)

Modern spelling of Gitksan name: gayda ts'uuts' (literally, 'hat of bird,' or 'bird-hat')

Editorial comments: Of this fungus Smith wrote only "[n]o medicine, no good, no story, grows on hemlock; same name because nearly hat, lamp + table."[33]

Lichens (Lichenized Fungi)

Parmeliaceae

(4) ?Cetraria pinastri (Scop.) S. Gray (Moonshine Cetraria), and possibly also other Cetraria spp.

Identified by Smith as: lichen from Scrub Pine (4) (15); Cetraria juniperina (S.) Ach. (Cetraria juniperina v. terrestris Schaur.) (from scrub pine, Kitwanga, B.C., August 23, 1925)

Smith's transcription of Gitksan name: clăănĭsĭs skinisht (clăănĭsĭs meaning lichen and skinisht yellow;[34] jack pine called skinisht)

Modern spelling of Gitksan name: hla'anisihl sginist (literally, 'the branch of lodgepole pine')[35]

Editorial comments: Smith, while in the Division of Anthropology at the Victoria Memorial Museum in Ottawa, sent two to several "yellow lichens" to G.K. Merrill of Rockland, Maine for

---

[32] Like the preceding fungus, this fungus is unidentified in botanical Latin terms. It is possible that it may represent the same species as the preceding entry, or a similar species.

[33] Based upon Smith's report, it seems possible that this Gitksan term may have a broader range of reference than its known Tsimshianic cognates that refer specifically to mushrooms.

[34] These are not the correct translations (see the modern spelling of the Gitksan term and the following footnote).

[35] This is probably a nonce form that simply refers to the branch of lodgepole pine, a substrate of the lichen in question, rather than to the lichen itself.

identification. In a letter dated December 2, 1925, Merrill identified Smith's collections (No. 4, 15) as "yellow lichen, <u>Cetraria juniperina</u> (S.) Ach." and "dark hair like-lichen, <u>Alectoria jubata</u> (S.) Myl." and the "bright yellow plant found on the ground in the region where (the) specimens originated is <u>Cetraria juniperina</u> v. <u>terrestris</u> Schaur."

Furthermore, according to Merrill:

"I do not know that the yellow plant is designated by a "common" name in this country. In Scandanavia [sic] where it is used by the inhabitants it receives a local name.

There is no reaction with KOH. or CaCl. but it yields a distinct yellow on maceration with water. I have no doubt but what ammoniacal maceration will produce either an intensified coloration, or produce some new shade. It seems not to be on the edible list. I have been told that the Indians employ the <u>Alectoria</u> when it is abundent [sic] enough, for food. The preparation consists of rolling the moistened mass of the lichen into a ball, and then baking it in the ground.[36]

The western Indians use <u>Letharia vulpina</u> a lichen of a similar color to the <u>Cetraria</u> for the manufacture of a dye.

There is another yellowish <u>Alectoria</u> from B.C. that afford[s] a distinct yellow on maceration with water. I have no information regarding its being used by the Indians."

Of the lichens, Smith wrote that:

Among the Gitksan, according to John Fowler August 27, 1925, this lichen was of no use; it was not used for making a dye for mountain goat[37] wool and no paint or medicine was made of it.

---

[36]The lichen referred to as <u>Alectoria</u> is actually one or more species of <u>Bryoria</u> known to have been cooked in earth ovens and used as food by several interior First Nations groups.

[37]Mountain goat (<u>Oreamnos</u> <u>americanus</u> [Blainville]) is known in Gitksan as matx. Note that the Gitksan animal names reported in this and subsequent footnotes are from Hindle and Rigsby (1973). The proposed zoological Latin identifications are based on known species distributions within Gitksan territory.

It seems likely that John Laknitz had heard of the use of the well known lichen that was used for a yellow dye[38] in north western North America but did not know that his lichen of Kitwanga was of another kind.

Among the Gitksan, according to John Laknitz, August 23, 1925, this lichen is found on scrub pines and on the rock in the mountains; there is only one kind of it, and he had heard that it was used for making a yellow dye for mountain goat wool. He said that he had not seen it used and that the yellow dye was not used for colouring such things as totem poles.

Stictaceae

(5) Lobaria spp. (Lung Lichens, Stictaceae), e.g., L. pulmonaria (L.) Hoffm.
Identified by Smith as: (31) "frog-blankets"
Smith's transcription of Gitksan name: gwillĕ ganao, "frog-blankets"
Modern spelling of Gitksan name: gwiilahl ganaa'w (literally, 'blankets of frog,' or 'frog-blankets')
Smith's transcription of Gitksan name: lū quĭla ganao, (lugoola ganao), "frog blanket" (lu meaning belongs to, quĭlā, blanket and ganao frog; so named because it is like the skin of a frog)[39]
Modern spelling of Gitksan name: luu gwiilahl ganaa'w ("there are frog-blankets inside")

Among the Gitksan, according to Luke Fowler, May 24, 1926, and Robert A. Sampare June 9, 1926, this plant grows on large willows or on cottonwoods and was of no use.

Bryophytes (Mosses and Their Relatives)

Sphagnidae

(6) Sphagnum spp., probably including S. angustifolium (Russ.) Tolf. (Poor-fen Sphagnum), S. capillaceum (Weiss) Schrank (Common Red Sphagnum), S. fuscum (Schimp.) Klinggr. (Common Brown Sphagnum), S. girgensohnii Russ. (Common Green Sphagnum), and S. squarrosum Crome (Shaggy Sphagnum)

---

[38]This lichen is likely Letharia vulpina (L.) Hue (wolf lichen).
[39]The Gitksan term translated as "frog" (ganaa'w) likely refers to various species of frogs and toads (Anura)—perhaps specifically to Bufo boreas (Baird and Girard) (western toad; Bufonidae)—rather than to any single true frog species.

Identified by Smith as: (145) <u>Sphagnum</u> sp. (Peat Moss) (from muskeg near Hazelton, B.C.)
Gitksan name: (no name recorded by Smith)[40]

Among the Gitksan, according to Bob Robinson, September 5, 1926, this moss was used for diapers for babies (see Figure 2).

---

[40]Hindle and Rigsby (1973:33) recorded the term umhl for "moss"—a term that likely includes reference to <u>Sphagnum</u> spp.

Figure 2. Sphagnum capillaceum (common red sphagnum)

(7) Unidentified Bryophyta?

Identified by Smith as: (116) unidentified "mosses" (two kinds)[41]

Gitksan name: (no name recorded by Smith)[42]

Among the Gitksan, according to Luke Fowler, May 24, 1926, these two mosses were of no use, but bears eat them.

Pteridophytes (Ferns and Their Relatives)

Dennstaedtiaceae and Dryopteridaceae

(8) ?<u>Athyrium filix-femina</u> (L.) Roth ssp. <u>cyclosorum</u> (Rupr.) C. Christ. in Hult. (Lady Fern; Dryopteridaceae) or possibly <u>Pteridium aquilinum</u> (L.) Kuhn ssp. <u>lanuginosum</u> (Bong.) Hult. and/or <u>P. aquilinum</u> (L.) Kuhn ssp. <u>latiusculum</u> (Desv.) C.N. Page (Bracken Fern; Dennstaedtiaceae, Hay-Scented Fern Family)

Identified by Smith as: unidentified fern "a fern, (collect)"[43]

Smith's transcription of Gitksan name: demk or demk°

Modern spelling of Gitksan name: demtx (WG), damtx (EG)

Among the Gitksan, according to Luke Fowler, May 24, 1926, the large round green rootstock of this fern, but not the rootlets were mashed together with bark of balsam fir[44] and devil's club, a little gum of scrub pine or spruce and root of Indian hellebore and the mass was warmed a little and applied to boils and ulcers. It caused them to come to a head. If this medicine was not applied they suppurated. It was also put on the chest for hemorrhage of the lungs, but sometimes the patient was too far gone. This fern was not used as a food (see Figure 3).[45]

---

[41] Smith did not identify these mosses, nor has it been possible to determine their identities from a search of voucher specimens collected by Smith and deposited at the National Herbarium of Canada (M.J. Shchepanek, pers. comm. 1991; P. Frank, pers. comm. 1995).

[42] Hindle and Rigsby (1973) recorded the term umhl for "moss."

[43] This species was neither observed nor collected by Smith.

[44] <u>Abies balsamea</u> (L.) Mill., an eastern species, is the true balsam fir although the three species of <u>Abies</u> that are native to British Columbia are often referred to as "balsam" or "balsam fir." Where "balsam" or "balsam fir" appears in the present document, some native species of fir is intended.

[45] Elsewhere, Smith (1929:48) reported the following additional medicinal information for "fern, species uncertain" (and the aforementioned combination of ingredients): "Also used for rheumatism, and as a plaster on the chest for hemorrhage of the lungs."

Figure 3. Athyrium filix-femina (lady fern)

Dryopteridaceae

(9) ?"Dryopteris austriaca complex," i.e., D. carthusiana (Vil.) H.P. Fuchs (Toothed Wood Fern) and/or D. expansa (K.B. Presl) Fraser-Jenkins & Jermy (Spiny or Spreading Wood Fern)[46]

Identified by Smith as: (37) Asplenium cyclosorum Rupr. (Lady Fern); syn.: Athyrium filix-femina (L.) Roth subsp. cyclosorum (Rupr.) C. Christ. in Hult. (Lady Fern)

Smith's transcription of Gitksan name: ak

Modern spelling of Gitksan name: ax

Among the Gitksan, according to Luke Fowler, May 24, 1926, this plant was one of the principal foods.[47]

Equisetaceae (Horsetail Family)

(10) Equisetum arvense L. (Common or Field Horsetail)

Identified by Smith as: (34a) Equisetum arvense L. (Horsetail Rush)

Smith's transcription of Gitksan name: mount (mow), "green meadow," mount means "file"

Modern spelling of Gitksan name: maawint, maawin[48]

Among the Gitksan, according to Luke Fowler, May 24, 1926, this plant was used as a file and sandpaper, in cleaning, polishing, and shining. It was used to polish arrow points made of wood of the wild rose, to smooth spoons made of maple wood, and large boxes (see Figure 4).

---

[46]This species complex and its relationship to First Nations' fern knowledge and use is discussed in Turner, et al. (1992).

[47]The fern identified in Gitksan as ak is likely wood fern, rather than the common lady fern. Elsewhere Smith referred to "a fern," "demk° in Gitksan" and associated species 37 with "a fern (to be collected)." In his discussion of skunk-cabbage, Smith stated that the reader should "see a fern," also referred to by name as demk°. Smith's specimen 37 was associated with the Gitksan name, ak, but the fern called demk or demk° was not identified in botanical Latin terms. This latter fern (possibly the species named above) apparently was combined with skunk-cabbage and other materials for use as a medicine.

[48]Editorial comments provided by the Sim'algax Working Group suggest that this term may actually refer to the following species, E. hyemale.

Figure 4. Equisetum arvense (common horsetail)

(11) <u>Equisetum</u> <u>hyemale</u> L. ssp. <u>affine</u> (Engelm.) Stone. (Scouring-rush)

Identified by Smith as: (34b) <u>Equisetum</u> <u>hyemale</u> L. (Branched Horsetail Rush)

Smith's transcription of Gitksan name: hish mount, "funny green meadow;" hish means "funny," mount (mow) means "file"

Modern spelling of Gitksan name: hismaawint ("it's a pretend/false-scouring-rush")[49]

Among the Gitksan, according to Luke Fowler, May 24, 1926, this plant was of no use.

Lycopodiaceae (Clubmoss Family)

(12) ?<u>Lycopodium</u> <u>complanatum</u> L. (Ground-cedar)

Identified by Smith as: (124) <u>Lycopodium</u> <u>complanatum</u> L. (Club Moss, Ground Pine) (from first terrace, Kitwanga, B.C., August 28, 1926)

Smith's transcription of Gitksan name: belana watsx (belana means "belt," watsx, "land otter")

Modern spelling of Gitksan name: bilena 'wats<u>x</u> (WG), bilana 'wats<u>x</u> (EG) (literally, 'belt of river otter,' or 'river otter-belt')

Among the Gitksan, according to Gus Sampare, September 30, 1926, this plant was of no use. It had no name known to Gus (see Figure 5).

---

[49]This may be a nonce form. The construction with initial his- and a final relativizer -(i)t conveys the meaning that the phenomenon signified by the stem verb or noun is not the real thing, and so hismaawint might be translated something like "it's a pretend-horsetail rush" or "it's a false-horsetail rush." Editorial comments provided by the Sim'alga<u>x</u> Working Group suggest that this term may actually refer to the preceding species, <u>E</u>. <u>arvense</u>, which would mean that the translation may actually mean "it's a pretend/false-field horsetail."

Figure 5. <u>Lycopodium</u> <u>complanatum</u> (ground-cedar)

(13) ?<u>Lycopodium dendroideum</u> Michx. (Ground-pine)

Identified by Smith as: (155) <u>Lycopodium obscurum</u> L. var. <u>dendroideum</u> Eaton (Club Moss)

Smith's transcription of Gitksan name: belana watsx, "land otter belt" (belana means "belt," watsx, "land otter")[50]

Smith's transcription of Gitksan name: belana watex, "feet land otter"

Modern spelling of Gitksan name: bilena 'wats<u>x</u> (WG), bilana 'wats<u>x</u> (EG) (literally, 'belt of river otter,' or 'river otter-belt')

Among the Gitksan, according to Bob Robinson, September 5, 1926, this plant was of no use.

Gymnosperms (Conifers and the Taxad, Western Yew)

Cupressaceae (Cypress Family)

(14) <u>Chamaecyparis nootkatensis</u> (D. Don in Lamb.) Spach (Yellow Cedar or Cypress, or Alaska Cedar or Cypress)

Identified by Smith as: <u>Chamaecyparis nootkatensis</u> (Lamb) Spach. (Yellow Cypress)

Gitksan name: (no name recorded by Smith)[51]

According to Luke Fowler, June 8, 1926, yellow cedar is not found in the Gitksan country.[52]

(15) ?<u>Juniperus communis</u> L. (Common or Ground Juniper)[53]

Identified by Smith as: (63) <u>Juniperus</u> sp.

---

[50]"Land otter," or river otter, is <u>Lontra canadensis</u> (Shreber). Among the Heńaaksiala and Haisla, <u>L. clavatum</u> L. (running clubmoss) is known as 'belt of river otter' (Compton 1993). The Gitksan terms associated by Smith with <u>Lycopodium</u> spp. here and in the following entry suggest that at least one—and possibly several—types of club moss were similarly known and named among the Gitksan.

[51]Hindle and Rigsby (1973:22) recorded the term wihl for "fir tree," although this term is cognate with other Tsimshianic terms that refer to yellow cedar.

[52]This tree does not occur within Gitksan territory but it is widely distributed all along the coast to the west (see Krajina et al. 1982) where it was accessible to the Gitksan who travelled along the Skeena River to the coast and may have obtained it near the mouth of the Skeena where yellow cedar extends down to sea level. Smith indicated that paddles were sometimes made of yellow cedar as well as from other types of wood such as western redcedar, Douglas maple and red alder.

[53]<u>Juniperus scopularum</u> Sarg. (Rocky Mountain juniper) also occurs in small populations near Kitwanga and above the railroad at Andimaul (J. Pojar, pers. comm. 1996), but this species has not been reported in association with Gitksan ethnobotanical studies (e.g., Gottesfeld and Johnson 1988).

Smith's transcription of Gitksan name: skannāknōk (skan means "plant," nāknōk, any good medicine that cures a cut at once but juniper was not applied to cuts)[54]

Modern spelling of Gitksan name: sgannaxnok (literally, 'naxnok plant')[55]

Among the Gitksan, according to Luke Fowler, June 8, 1926, the juniper was used for small bows, it being too small for large ones. Sometimes it was used for good firewood, but it is small and there is not much of it. It was prized as a great internal medicine for which the whole plant, including the roots and berries, were boiled for a whole day. When cool the decoction was drunk for many ailments, hemorrhage at the mouth, and kidney trouble. It cleans one, being a purgative and diuretic, and makes one strong (see Figure 6).[56]

---

[54] The word reported here by Smith to mean "any good medicine..." (i.e., naxnok) actually refers to a kind of spirit-being that confers power on human owners As an intransitive verb, naxnok means "be clever, knowledgeable."

[55] In both the Coast and Southern Tsimshian languages similar terms incorporating cognates of naxnok also may, perhaps, apply to juniper which has had ritual or spiritual uses among shamans and others from some Central and North Coast First Nations (cf. Compton 1993). Gottesfeld and Anderson (1988:25, 26) have reported the Gitksan term laxsa laxnok that they say refers to any juniper or "to a specific ecotype of juniper growing in rocky places in the mountains." This term seems to represent a misspelling of laxsa naxnok meaning naxnok 'brush,' where the "brush" may be in reference to conifer boughs. The term sgan, when followed by another term for a plant part or characteristic, indicates reference to the plant in question in its entirety. For the purposes of this document, this term will be translated as 'plant,' although the translations 'shrub' or 'tree' might be appropriate in some cases and although the term may have a more basic literal meaning. (See the Discussion, Summary & Conclusions section for further discussion of the Gitksan 'plant' term within the context of Gitksan botanical nomenclature and classification.)

[56] Elsewhere, Smith (1929:49) identified *J. communis* as a medicinal plant among the Nuxalk and Carrier with uses very similar to those reported for the Gitksan.

Figure 6. Juniperus communis (common juniper)

(16) Thuja plicata Donn ex D. Don in Lamb. (Western Redcedar)

Identified by Smith as: Thuja plicata Donn. (Red Cedar)

Smith's transcription of Gitksan name: āmmedĕtl (āmmedell) (ām means "good," hadetl means "red cedar bark" after it is peeled and rolled up)

Modern spelling of Gitksan name: amhat'e'l (WG), amhat'a'l (EG) (literally, 'good for-inner redcedar bark'—hat'e'l [WG], hat'a'l [EG], 'inner redcedar bark')

Among the Gitksan, according to Luke Fowler, May 24, 1926, redcedar trees were used for totem poles, house frames and ridges, and large canoes, although there were not many cedar trees of sufficient size for canoes and most of them were made of cottonwood. Young cedar trees were used for wedges and for handles for spear points made of hemlock knots used for bears and men. For this purpose they were lighter than those made of big saskatoon bushes. Cedar poles were used for the wings of salmon[57] traps. Being easy to split, cedar was used for boards for all the houses and bent-wood boxes. Split fine, it was used for fish traps, but was rather too light. The wood was also used for paddles and some of the dugout dishes. It was too light for the long handles of salmon gaffs and if used for that purpose the gaff would be carried out of control by the current of the swift waters. The limbs were used as rope, being crushed and used singly, or neatly twisted into two or three strand rope, or braided. Those from stunted crowded trees were used for tying bridge timbers together. The bark was used for house roofs because it never burns quickly. The inner bark was used for slow matches used to carry fire from one camp to another, mats, baskets, and head and neck rings, also, where the water was not too swift, to lash the parts of salmon traps together. The roots were not used, although those of the spruce were. The tree produced no medicine. The inner bark was used in its natural colour, and dyed both black and red. It was dyed black by being soaked in blue silver waters, or red iron waters and was coloured red with a decoction of red alder bark.

Among the Gitksan, according to Luke Fowler, June 8, 1926, redcedar disappears from the Skeena Valley forty miles above Hazelton, B.C. It was used for lumber and in its absence balsam fir was used. The bark was used for roofs (see Figure 7).

Editorial comments: According to Smith, fire drill hearths were made of cedar or willow and cedar mats were used to cover earth ovens used for cooking lupine roots.

---

[57] Several species of salmon were known to and used by the Gitksan. In general, salmon (Oncorhynchus spp.) are referred to in Gitksan as hun.

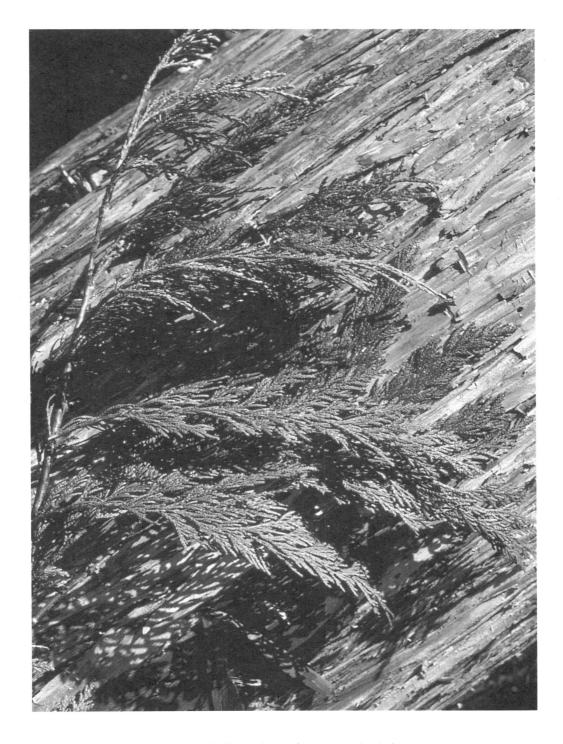

Figure 7. Thuja plicata (western redcedar)

Pinaceae (Pine Family)

(17) Abies amabilis (Dougl. ex Loud.) Forbes (Amabilis or Pacific Silver Fir) and A. lasiocarpa (Hook.) Nutt. (Subalpine or Alpine Fir)
Identified by Smith as: Abies sp. (Balsam Fir)[58]
Smith's transcription of Gitksan name: hăwāks (hawox)
Modern spelling of Gitksan name: hoo'oxs[59]

Among the Gitksan, according to Luke Fowler, June 8, 1926, the wood of balsam fir was used for lumber in the absence of redcedar. It was good firewood. Wood of young balsam firs was used for making snowshoes. The bark was used for good roofs. The juicy inner bark was scraped from the trunk of the tree, with a bone scraper after the outer bark had been removed, and was taken for constipation in June. As a medicine for many bad ailments, the gum of the bark blisters or the young cones, which can be obtained in August sliced across and mashed, was taken internally as a purgative and diuretic for consumption (tuberculosis) and gonorrhea. It was applied externally to cuts and sores, especially the sores of gonorrhea. The bark of balsam fir and some other materials were mixed with the fern, demk° (demtx [WG], damtx [EG]), for use as a medicine.[60] The roots were of no use (see Figures 8 and 9).

---

[58]Elsewhere,"balsam fir" was identified by Smith (1929:48, 50) as "Abies grandis Lindl. (White Fir, Balsam Fir)" (i.e., A. grandis [Dougl. ex D. Don in Lamb.] Lindl., grand fir), a species that does not occur as far north as Gitksan territory. All of the information associated by Smith with "balsam fir" must refer to amabilis fir and possibly also to subalpine fir, species that do occur within Gitksan territory. Abies lasiocarpa is found in valley bottoms and on lower slopes while A. amabilis is abundant at middle elevations and in some side drainages such as Mudflat Creek and Boulder Creek where the trees have been protected from fire (J. Pojar, pers. comm. 1996).

[59]The form amhoo'oxs has also been recorded for this species.

[60]Guédon (1973) has also reported that the pitch from the lumps in balsam fir bark was used used for colds and arthritis and, when combined with eulachon grease, for rheumatism. The pitch is said to be good for any kind of infection. The Gitksan ethnobotanical data reported by Guédon correspond to a set of 30 ethnobotanical voucher specimens collected by Guédon and Rosalind Whalley during June 1973 at a low mountain near Kathleen Lake approximately 60 miles from Hazelton, Kispiox and Glen Vowell. These specimens were subsequently identified by Brian D. Compton and provided to the Canadian Museum of Civilization where they have been assigned the catalog number LH 996.16.1 in the Living History Collection.

Figure 8. Abies amabilis (amabilis fir)

Figure 9. Abies lasiocarpa (subalpine fir)

(18) <u>Picea</u> x <u>lutzii</u> Little (Hybrid Sitka Spruce)[61]
Identified by Smith as: <u>Picea sitchensis</u> Carr (Sitka Spruce)
Smith's transcription of Gitksan name: seaks (sayox)
Modern spelling of Gitksan name: see<u>k</u>s

Among the Gitksan, according to Luke Fowler, May 24, 1926, [hybrid] Sitka spruce was used for firewood, and for the long handles of salmon gaffs. It made lighter handles than those sometimes made of scrub pine. Spruce was occasionally used for the wings of salmon traps. It was not good for paddles. The bark was used for house roofs in the absence of redcedar. Scrapings of the inner side of the bark and the peeled tree trunk were eaten, but were not dried or made into dry cakes. The gum extracted from the wood by boiling in water, was added to eulachon[62] or salmon oil or bear grease, groundhog (or woodchuck)[63] fat or, in fact, any edible oil, and was drunk before meals for consumption. In recent times it was taken with lard. The roots were woven into rain proof hats and basket kettles in which meat and salmon were boiled by adding hot stones. After the water was scraped out of the roots, they were used for tying the parts of salmon traps together, especially where the water was swift (see Figure 10).

<u>Editorial comments</u>: Spruce gum, when mixed with a fern identified only as demk° (demtx [WG], damtx [EG]), Indian hellebore, and other plants, was used as medicine. The gum obtained from the boiled wood was drunk for measles, as well as consumption. Elsewhere Smith (1929:52) reported the following medicinal information for [hybrid] Sitka spruce: "Twigs [of hybrid Sitka spruce] bearing both leaves and bark boiled with entire roots of soopolallie (<u>Shepherdia</u> <u>canadensis</u> Nutt.); one cupful of the strong decoction taken internally three times a day for rheumatism."[64]

---

[61] This is a hybrid between <u>P. glauca</u> (Moench) Voss (white spruce) and <u>P. sitchensis</u> (Bong.) Carr. (Sitka spruce) (Douglas et al. 1989:7). Other than black spruce (<u>P. mariana</u> [P. Mill.] B.S.P.), most, if not all, of the spruce in Gitksan territory is hybrid Sitka spruce (J. Pojar, pers. comm. 1996). If not otherwise noted, any reference in the text to "spruce" denotes hybrid Sitka spruce.
[62] Eulachon, <u>Thaleichthys</u> <u>pacificus</u> (Richardson) is known in Gitksan as saak, a loanword from Tlingit.
[63] Groundhog or woodchuck (<u>Marmota</u> <u>monax</u> [L.]) is known in Gitksan as gwiikw.
[64] Guédon (1973) has also reported that hybrid Sitka spruce pitch was applied to boils and infections to draw out the pus. Sometimes, a sliver of redcedar bark was pushed into the boil and pitch was applied so that when the sliver was removed, the pus would escape. Hybrid Sitka spruce bark was also used as an antiseptic splint used over an application of yellow pond lily rhizomes and "wild rhubarb" (cow-parsnip) roots for treating broken bones. The boiled bark was considered to be good as a tuberculosis medicine.

Figure 10. Picea x lutzii (hybrid Sitka spruce)

(19) Pinus contorta Dougl. ex Loud. var. latifolia Engelm. (Lodgepole Pine)
Identified by Smith as: Pinus contorta Dougl. (Scrub Pine)
Smith's transcription of Gitksan name: skĭn`ĭsht
Modern spelling of Gitksan name: sginist[65]

Among the Gitksan, according to Luke Fowler, June 8, 1926, clear scrub pine timber was split for salmon traps. Poles of this tree were sometimes used for poling canoes, as any kind of tree might be used for that purpose, although poles of small young hemlocks were considered the best. The wood, especially the resinous parts, was considered good for fires. The resinous wood was also used for torches. The inner bark was used for food and as a blood purifier and purgative. Both resinous shavings and young needles were used for a purgative and diuretic. Scrub pine wood was not used for paddles and the roots were not employed in making baskets.

Torches were made by tying about twelve pounds of strips of the resinous wood into a bundle and such torches were used in spearing fish and to light the way in travelling. Scrapings of the inner bark taken with a bone scraper after the outer bark was peeled off were called gānĭh [ganhix] and were eaten as food. They caused purging in from half an hour to an hour and made one hungry. Shavings of the yellow resinous timber after the bark was taken off were boiled, the decoction put in oil and drunk as a purgative and diuretic for many bad ailments. It tastes like water; was used for gonorrhea; and was somewhat helpful in consumption. Young needles were plucked in June and eaten as a purgative and diuretic (see Figure 11).

Editorial comments: According to Smith, gum from scrub pine or spruce was mixed with other ingredients for use as a medicine for boils and ulcers (see Lysichiton americanum). A lichen said to be collected from scrub pine (probably Letharia vulpina, wolf lichen), was used as a source of yellow dye—by western groups but not specifically the Gitksan.[66]

---

[65]This word is related historically by derivation from sgan, 'pitch.'
[66]Very little Letharia vulpina occurs within Gitksan territory, except for its occasional occurrence on Juniperus scopularum (Rocky Mountain juniper), which itself is rare there (J. Pojar, pers. comm. 1996).

Figure 11. Pinus contorta (lodgepole pine)

(20) Tsuga heterophylla (Raf.) Sarg. (Western Hemlock)
Identified by Smith as: Tsuga heterophylla Sarg. (Western Hemlock)
Smith's transcription of Gitksan name: gico·ᵘ
Modern spelling of Gitksan name: giikw[67]

Among the Gitksan, according to Luke Fowler, May 24, 1926, hemlock was used for firewood. The young trees were used for wedges but the knots were considered too hard for wedges, although the Nuxalk made their wedges out of them. Spear points made of the knots were used for bears and men. These had handles made of a large saskatoon bush or of a young redcedar which is still lighter. The very young trees were not split for salmon traps or spear points.

For food the outer bark was taken off and the inner bark scraped from it. This was eaten with eulachon or salmon oil, bear grease or groundhog grease, in fact any grease. It was very good. It was also made into dry cakes for winter use (much as among the Nuxalk. For this purpose stones were placed on a fire built in a hole in the ground. After the fire had burned down and the stones were very hot the inner bark of the hemlock was placed on top of them and left to bake until morning. Then the baked bark was pounded in a large mortar for many hours after which it was formed into thin cakes and dried on a rack in the sun. It was not dried over a fire, because that would make it taste like smoked wood. To serve, it was soaked in water and eaten with any kind of grease. The tree produced no remedies (see Figure 12).[68]

Editorial comments: Smith also stated that poles of small young hemlocks were considered the best to pole canoes upriver.

---

[67] The form amgiikw has also been recorded for this species. The literal meaning of giikw is unknown, but the homophonous transitive verb root means "buy, choose."

[68] Elsewhere, Smith (1929:51) stated that neither T. heterophylla nor T. mertensiana (Bong.) Carr. (mountain hemlock) was used by the Gitksan for medicine.

Figure 12. Tsuga heterophylla (western hemlock)

Taxaceae (Yew Family)

CAUTION—THE FOLLOWING SPECIES IS TOXIC!

(21) <u>Taxus</u> <u>brevifolia</u> Nutt. (Western or Pacific Yew)
Identified by Smith as: <u>Taxus</u> <u>brevifolia</u> Nutt. (Yew)
Smith's transcription of Gitksan name: hāgtuk^u or tuk^u ("shooting timber")
Modern spelling of Gitksan name: haxwdakw or xwdakw[69]

<u>Editorial comments</u>: Smith wrote that the wood of western yew was used for adze handles. Apparently, according to Smith's comments, this tree did not grow in Gitksan territory, but the wood could be bought at Fort Simpson. It was strong and never broke. The Gitksan name means "shooting timber" (see Figure 13).[70]

---

[69] Xwdakw is the singular form of the intransitive verb "shoot," while haxwdakw ("bow") is a derived form that might be glossed as 'instrument for shooting.'
[70] This name may refer to the use of this wood for making bows, a use not reported here but which would be consistent with the use of yew elsewhere in the area (Compton 1993).

Figure 13. Taxus brevifolia (western yew)

Angiosperms (Flowering Plants), Dicotyledons

Aceraceae (Maple Family)

(22) <u>Acer glabrum</u> Torr. var. <u>douglasii</u> (Hook.) Dippel (Douglas or Rocky Mountain Maple)
Identified by Smith as: (35) <u>Acer glabrum</u> Torr. (Maple)
Smith's transcription of Gitksan name: gäs (gaws) or gäst
Modern spelling of Gitksan name: k'oox̱st

Among the Gitksan, according to Luke Fowler, May 24, 1926, there is only one kind of maple in the Gitksan country. The inner bark was used to make woven baskets, large woven pack sacks, and woven mats. It was dyed black by being put in a certain mud, red by being soaked in another kind of mud, or by being put in red alder dye. the wood was used for spoons, snowshoes, shaman's spherical rattles, and chief's bird-shaped rattles, but not for masks. Bunches of the seeds were shaken by children as toy rattles. The roots and leaves were of no use.

<u>Editorial comments</u>: Smith also wrote that maple wood was used for soopolallie spoons and arrows which were smoothed with common horsetail. Maple and red alder were said to be the best woods for paddles (see Figure 14).

Figure 14. Acer glabrum (Douglas maple)

Apiaceae (syn.: Umbelliferae, Parsley Family)

(23) *Angelica genuflexa* Nutt. (Kneeling Angelica)

Identified by Smith as: (114) (136) (143) *Angelica genuflexa* Nutt. (Angelica, Wild Parsnip)

Smith's transcription of Gitksan name: hammok ganao, "frog parsnip" (hammok meaning "parsnip" and ganao meaning "frog")[71]

Smith's transcription of Gitksan name: clamok ganao, "frog rhubarb" (hammok means "rhubarb" and ganao "frog")[72]

Smith's transcription of Gitksan name: hămăk ganao meaning the seeds, and ganao meaning frog

Modern spelling of Gitksan name: ha'mookhl ganaa'w ("frog cow-parsnip," literally, 'sucker/sucking-tube of frog,' or 'frog-sucker/sucking-tube')[73]

Smith's transcription of Gitksan name: ēsim ganaootus or ēsimsknao otus (ēsim means "to urinate," sknao or ganao means "frog")

Modern spelling of Gitksan name: isimganaa'wtxws (literally, 'to urinate like a frog')

Among the Gitksan, according to Bob Robinson, September 5 and October 3, 1926, the roots were held in the mouths of gamblers and they spit on their hands. They had to keep close to the fire and away from their wives for twenty days before gambling.

According to John Fowler, September 16, 1925, the hollow stalks were used for blow-guns in which boys shot choke cherry pits. It was said the seeds were eaten by frogs.

Editorial comments: Smith seems to have been somewhat uncertain about whether this species or cow-parsnip was used by gamblers as stated above. However, he did not conclusively state that cow-parsnip was used for this purpose and, in one case in the original manuscript, he crossed out text referring to the use of cow-parsnip by gamblers. This confusion probably resulted from the similar appearance of these two plants and their similar Gitksan names.

---

[71] Here, "parsnip" is used to refer either to kneeling angelica or *Heracleum lanatum* (cow-parsnip). This should not be confused with the European species, *Pastinaca sativa* L. (common or wild parsnip), which has been introduced to, and is naturalized in, British Columbia.

[72] The "rhubarb" referred to here is *Heracleum lanatum* (cow-parsnip), rather than the domesticated garden variety, *Rheum rhabarbarum* L.

[73] Ha'mook is the conventional name for "rhubarb" and "cow-parsnip." It is derived from the intransitive verb 'mook, "suck," and might be glossed literally as "instrument for sucking" or "sucking-tube" (e.g., as in a straw—these plants have hollow stems). The =hl connective that Rigsby has inserted here marks the following noun as a common noun. It is more likely that Smith would have missed it, than if his consultant had used the -s proper noun connective. Ha'mooks Ganaa'w would be 'sucker/sucking-tube of Frog' or 'Frog's sucker/sucking-tube,' while ha'mookhl ganaa'w is 'sucker/sucking-tube of frog' or 'frog-sucker/sucking-tube.'

Unlike cow-parsnip, the stems of kneeling angelica were not generally eaten. However, according to Luke Fowler, if one did eat this plant it would prevent that person from being smelled by bears when hunting. The hollow stems were commonly used to drink water, as were the shorter hollow stems of cow-parsnip.

Elsewhere, Smith (1929:61) reported the following medicinal information for kneeling angelica: "Roots well boiled with twigs of squashberry (<u>Viburnum pauciflorum</u> Raf.)[74] from which the bark had not been removed, and decoction taken internally for headache and weak eyes." He also attributed similar comments to "135," i.e., a plant he identified as agrimony: "Among the Gitksan, according to Luke Fowler, May 24, 1926, the roots of this plant were used for medicine. They were dried for use out of season. A decoction was made of the roots well boiled with twigs of the highbush cranberry from which the bark had not been removed. This was drunk for headache and weak eyes."

CAUTION—THE FOLLOWING SPECIES IS CAPABLE OF CAUSING DERMATITIS!

(24) <u>Heracleum lanatum</u> Michx. (Cow-parsnip)
Identified by Smith as: <u>Heracleum lanatum</u> Mich. (Cow Parsnip) (from Kitwanga-Kitwankool road, 1925)
Smith's transcription of Gitksan name: hammok (Fowler and Robinson); the leaf stalks are called beanst; the cylindrical stalks hōk (Luke Fowler)[75]
Smith's transcription of Gitksan name: hămäwk (hämäk is tube which water = pipe)
Modern spelling of Gitksan name: ha'moo<u>k</u> (literally, 'instrument for sucking')
Modern spelling of Gitksan name: p'iinst (petioles)

According to Luke Fowler, May 24, 1926, and Bob Robinson, October 3, 1926, the inside of the tender new shoots are eaten (as they are among the Nuxalk. Luke Fowler also stated that the roots were mashed and tied on rheumatic swellings, boils and other swellings.[76] The hollow stalks were used, in the absence of a hollow bone, for a drinking tube by girls during

---

[74]This is an obsolete synonym for <u>Viburnum edule</u> (Michx.) Raf. (highbush cranberry).

[75]The term reported by Smith for "leaf stalks" seems to correspond to Coast and Southern Tsimshian p'iins, which refers to the petioles (leaf stalks) of cow-parsnip. Gottesfeld and Anderson (1988) have reported huuxk to refer to the "flower bud stalk" of cow-parsnip.

[76]Guédon (1973) has also reported that the roots of cow-parsnip were combined and cooked with hybrid Sitka spruce pitch to treat boils and infections. Cow-parsnip roots were mixed with salt and used to treat broken bones.

the rites of puberty. It was thought that a girl who looked at a mountain at that time would become blind so they were kept in a lodge where they would not see out (see Figure 15).

Editorial comments: According to Smith, young boys shot choke cherry pits from blow guns made from the dried stems of cow-parsnip as well as of kneeling angelica.

Figure 15. Heracleum lanatum (cow-parsnip)

(25) Osmorhiza chilensis H. & A. (Mountain Sweet-cicely)

Identified by Smith as: (84) Osmorrhiza divaricata Nutt. (Sweet Cicely)

Smith's transcription of Gitksan name: hīslēuqōt (hislayuqut), his meaning "similar," lēok,

"spreading dogbane," the plant used for fibre.[77]

Modern spelling of Gitksan name: his-(unrecognizable, perhaps spreading dogbane)-xwit ("it's a pretend/false-[undefined]")[78]

Among the Gitksan, according to Bob Robinson, July 24, 1926, this plant was of no use and there was no story about it. The seeds stuck in the clothing of long ago.

Apocynaceae (Dogbane Family)

(26) Apocynum androsaemifolium L. (Spreading Dogbane)

Identified by Smith as: (242) Apocynum sp. (Spreading Dogbane)

Smith's transcription of Gitksan name: lēok, lē ok, lā ok

Modern spelling of Gitksan name: (unrecognizable, possibly leekw)[79]

Among the Gitksan, according to Luke Fowler, May 24, 1926, the fibre of this plant was the best used by the Gitksan for cord. The plant had no other use. The cord was employed for fish nets, for the warp, or short crosswise element, of pack-straps, and for fox,[80] lynx,[81] and groundhog snares. Such cord is very strong, so much so that it will cut one's flesh.

The plant was first dried, and then the bark was split with an awl and peeled from the stalk and the fibre twisted into cord by means of a spindle revolved by rolling it on either the thigh or the calf of the leg (see Figure 16).[82]

---

[77] Smith did not offer any explanation for why A. dioicus and Apocynum androsaemifolium were associated in nomenclatural terms.

[78] This term may be analyzed as his-(spreading dogbane?)-xwit, "it's a pretend/false-(spreading dogbane?)-(undefined)." This is probably a nonce form and is the same term as recorded for Aruncus dioicus.

[79] No other Tsimshianic or North Wakashan terms are recorded for this species and the term shown here does not correspond to the Nuxalk or Carrier terms for this species. The term posited here—leekw—has not been verified by contemporary Gitksan speakers but Guédon (1973) reported that this term was the Gitksan name for Apocynum spp.

[80] Red fox, Vulpes vulpes (L.), is known in Gitksan as t'ak'aluts or k̲ ala ẅaa'a.

[81] Lynx (Lynx canadensis Kerr) is known in Gitksan as weex.

[82] Smith elsewhere commented that this plant was spun on the thigh, not with a spindle. Two single strands were produced separately, then combined by rolling them backwards on the thigh together, so that they wrapped around each other to produce a two-ply strand.

Figure 16. Apocynum androsaemifolium (spreading dogbane)

Araliaceae (Ginseng Family)

(27) Aralia nudicaulis L. (Wild Sarsaparilla)
Identified by Smith as: (46) Aralia nudicaulis L. (Sarsaparilla) (from near Hazelton, B.C., 1926)
Smith's transcription of Gitksan name: maet^{u}smex (myt^{u}Smek), "bear berries," mae meaning "berry" and smex meaning "bear"
Modern spelling of Gitksan name: maa'ytxwhl smex (WG), maa'ytxwhl smax (EG) (literally, 'berries gathered by a black bear,' or 'the berries black bears gather')

According to Luke Fowler, May 24, 1926, black and grizzly bears[83] eat the berries.

CAUTION—THE FOLLOWING SPECIES IS POTENTIALLY HAZARDOUS (MAY CAUSE MECHANICAL INJURY AND IRRITATION)!

(28) Oplopanax horridus (Smith) Miq. (Devil's club)
Identified by Smith as: Fatsia horrida B. & H. (Devil's Club)
Gitksan name: (no name recorded by Smith)[84]

Among the Gitksan, according to Luke Fowler, May 24, 1926, devil's club was considered the best of all remedies and all the people knew of it. If it was taken for three or four months broken bones would knit.

Among the Gitksan, according to Luke Fowler, July 8, 1926, a decoction of the bark of this plant was used as a purgative in the treatment of gonorrhea (see Figure 17).[85]

---

[83] The black bear (Ursus americanus Pallas) is known in Gitksan as sim smax while grizzly bear (U. arctos L.) is called lik'insxw.

[84] Hindle and Rigsby (1973:18) recorded the term hu'ums or w'ums for devil's-club.

[85] Elsewhere, Smith (1929:62) reported the following medicinal information for devil's-club: "A decoction used as a purgative in the treatment of gonorrhea." "Boiled, together with entire plant of squashberry (Viburnum pauciflorum Raf.) and the decoction taken internally as a diuretic and purgative, for strangury or any sickness. Used continuously for rupture." Guédon (1973) has also reported that a milky infusion of devil's-club inner bark scrapings was made by soaking the bark in water and squeezing it to extract the medicine. This infusion was used to treat boils and other types of skin infections and rashes.

Figure 17. Oplopanax horridus (devil's club)

Asteraceae (syn.: Compositae, Aster or Composite Family)

(29) <u>Achillea</u> <u>millefolium</u> L. (Yarrow)

Identified by Smith as: (27) <u>Achillea</u> <u>lanulosa</u> Nutt., or <u>A</u>. <u>millefolium</u> L. (Yarrow)

Smith's transcription of Gitksan name: snīgāntu (skneegan too)[86]

Modern spelling of Gitksan name: (unrecognizable)

Among the Gitksan, according to Luke Fowler, May 24, 1926, all the plant except the roots was boiled and the decoction was gargled for sore throat.[87] Old plants were not used for this purpose, but young ones were taken in June until the middle of July. They were not dried for use out of season. Luke Fowler said he had heard that in the winter the dead stalks marked the place where the roots might be found.

Among the Gitksan, according to Abraham Fowler, August 9, 1925, this plant was of no use (see Figure 18).

---

[86]This term is similar to, and possibly is the same as, the term Smith reported for the following species, pearly everlasting (<u>Anaphalis</u> <u>margaritacea</u>).

[87]Guédon (1973) has also reported that yarrow was crushed and mixed with pitch for use as a poultice.

Figure 18. Achillea millefolium (yarrow)

(30) <u>Anaphalis</u> <u>margaritacea</u> (L.) Benth. & Hook. f. ex C.B. Clarke (Pearly Everlasting)
Identified by Smith as: (76) <u>Anaphalis</u> <u>margaritacea</u> Benth. (Pearly Everlasting) (from Hazelton, B.C., July 18, 1926)
Smith's transcription of Gitksan name: sninānt^u
Modern spelling of Gitksan name: (unrecognizable)

Among the Gitksan, according to Bob Robinson, July 18, 1926, this plant was used only for the flowers which, keeping perhaps two years, were long ago placed on coffin boxes and graves.

(31) <u>Artemisia</u> <u>michauxiana</u> Bess. in Hook. (Michaux's Mugwort)
Identified by Smith as: (123) <u>Artemisia</u> <u>discolor</u> Dougl. (Green Wormwood) (from bottom terrace Kitwanga, B.C., August 21, 1926)
Smith's transcription of Gitksan name: his gān tŭit (his means "nearly," gāntŭit, "stick") (tooit)
Modern spelling of Gitksan name: hisgantxwit ("it's a pretend/false-stick")[88]

Among the Gitksan, according to Gus Sampare, September 30, 1926, and Bob Robinson, October 3, 1926, this plant was of no use.

(32) <u>Aster</u> <u>conspicuus</u> Lindl. in Hook. (Showy Aster)
Identified by Smith as: (77) <u>Aster</u> <u>conspicuus</u> Lindl. (Aster) (from Hazelton, B.C., July 18, 1926)
Smith's transcription of Gitksan name: skanmīus, skan mīus or skan mī use, skan meaning "plant" and mīus meaning "good smelling"
Modern spelling of Gitksan name: sganmiyuxws (literally, 'smell good plant,' or 'be good-smelling plant')

Among the Gitksan, according to Bob Robinson, July 18, 1926, this plant was of no use.

(33) <u>Lactuca</u> <u>biennis</u> (Moench) Fern. (Tall Blue Lettuce)
Identified by Smith as: (89) <u>Lactuca</u> <u>spicata</u> (Lam.) Hitchc. (Tall Lettuce)

---

[88]This is probably a nonce form and is the same term given by Smith for <u>Spiraea</u> <u>douglasii</u> and Guédon (1973) for "white pussytoes, <u>Antennaria</u> sp."

Smith's transcription of Gitksan name: skanqats or skan qats, skan meaning "plant" and quats
meaning "excrement," so named because of its bad odour
Modern spelling of Gitksan name: sgankw'ats (literally, 'excrement plant')

Among the Gitksan, according to Bob Robinson, July 24, 1926, this plant was of no use and had no story about it.

(34) †Sonchus arvensis L. (Sow-thistle)[89]
Identified by Smith as: (128) (129) Sonchus arvensis L. var. maritimus (L.) Wg. (Field Sow
Thistle) (from railway between Nash and Andimaul, B.C., September 1, 1926)
Smith's transcription of Gitksan name: gan lim lak skan ist ("grows on the mountain," so
named because it is common on the mountains; Skan meaning "plant")[90]
Modern spelling of Gitksan name: ganlaxsga'nist (literally, 'mountain plant')

Among the Gitksan, according to bob Robinson, October 3, 1926, this plant was of no use.

Betulaceae (Birch Family)

(35) Alnus rubra Bong. (Red Alder) and possibly also A. tenuifolia Nutt. (Mountain Alder)[91]
Identified by Smith as: (20) Alnus rubra Bong. (Red Alder) (Seen at Kitwanga, B.C., August 9, 1925.)
Smith's transcription of Gitksan name: am loose, "good" loose = "neck ring"
Modern spelling of Gitksan name: amluux (literally, 'good for-luux'), luux (literally, 'red
alder,' 'neck ring')[92]

Among the Gitksan, according to Abraham Fowler, the bark of this tree was boiled for a red dye (see Figure 19).

---

[89] A dagger (†) preceding a Latin name is used to denote an introduced species.

[90] Although Smith here has referred to sgan, meaning "plant," this form does not seem to occur in the term he recorded for sow-thistle.

[91] Based upon the comments of some contemporary Gitksan elders it seems likely that amluux actually refers to both A. rubra and A. tenuifolia (L. Gottesfeld, pers. comm. 1995). Alnus rubra occurs as far inland as Cedarvale but between Cedarvale and Kitseguecla there may be hybridization between A. rubra and A. tenuifolia. Most of the alder in Gitksan territory is A. tenuifolia (J. Pojar, pers. comm. 1996).

[92] Luux refers both to red alder (and mountain alder) as well as to a special type of neck ring worn by chiefs and shamans and composed of redcedar inner bark dyed red with a decoction of red alder bark.

Editorial comments: Smith also stated that this red dye was used to colour inner redcedar bark used for neck rings. According to Luke Fowler, May 24, 1926, cedar neck rings were "not so good" if they were not dyed with red alder bark but "the red makes it fine." For use as a dye agent, the bark was scraped from the wood and put in water to soak. This was used to dye maple wood for woven baskets and sometimes used to colour wool and to soften hides.

Red alder and maple wood were considered to be the best for paddles but it was too soft for wedges and was not used for canoe bailers. Red alder wood was also used for adze handles as was black hawthorn wood. It was also considered good for firewood and was used for masks. Red alder branches were used as an undermat upon which Gitksan women cleaned salmon.

Elsewhere Smith (1929:55) reported the following medicinal information for red alder: "Bark and roots boiled for about six hours and the decoction drunk in the morning for a cough. Bark from the stem, but not from the roots, scraped, mixed with water, and the infusion taken internally, as an emetic and purgative, for headache and many other maladies." This was also said to be used in the same way as red elder berry root medicine, which was used as a purgative.

Figure 19. Alnus rubra (red alder)

(36) Alnus tenuifolia Nutt. (Mountain Alder) or possibly A.. crispa (Ait.) (Green or Sitka Alder)[93]

Identified by Smith as: (7) Alnus tenuifolia Nutt. (Mountain Alder) (from Kitwanga, B.C., August 23, 1925)

Smith's transcription of Gitksan name: gĭĭsht or gĭsht, the catkins both pistillate and staminate being called mĕgagĭĭst, the leaves yensĭgĭĭst, yens meaning "leaves"

Modern spelling of Gitksan name: giist, meega giist (literally, 'cone [catkins] of giist'),[94] 'yensa giist (literally, 'leaves of giist')

John Fowler said many of these trees grow up on the mountains but only a few in the valley of the Skeena River near Kitwanga.[95] He believed the staminate catkins to be young pistillate catkins.

Among the Gitksan, according to John Fowler, August 27, 1925, the bark and roots of this tree were used for a cough medicine, the pistillate catkins for a physic, but the staminate catkins were of no use. Abraham Fowler, August 9, said the pistillate catkins and shavings were used for a gonorrhea medicine and the wood was used as material for spoons while John said it was used only for firewood. The branches John said were not used for withes (although it is well known that they are among the Nuxalk. Red dye was not made from this tree but was from the red alder.

For a cough medicine the bark and roots were boiled for about six hours and the decoction was taken in the morning. The physic was made by crushing the pistillate catkins. The mass was eaten when one was not strong but sleepy and thin. The pistillate catkins were not boiled. For gonorrhea, according to Abraham Fowler, August 9th, the pistillate catkins and shavings were taken raw or they were boiled and the decoction was taken three times a day. It was a diuretic, he said, and would effect a cure in one week.[96]

(37) Betula papyrifera Marsh. var. papyrifera (Paper, White or Canoe Birch)

---

[93] It seems likely that giist actually refers to A. crispa (one or both of two supspecies if both occur within Gitksan territory) while A. tenuifolia may be referred to as amluux along with A. rubra (L. Gottesfeld, pers. comm. 1995).

[94] This term is similar to meek, 'pine cone,' and seems to indicate a perceived resemblance between pine cones and the mature, woody, cone-like female inflorescences of Alnus spp.

[95] This information suggests that Mr. Fowler was referring to A. crispa, rather than to A. tenuifolia. It is unclear to which alder species the other information provided by Mr. Fowler refers but possibly A. tenuifolia is meant.

[96] Elsewhere Smith indicated that the withes of this plant may have been twisted for use in tying (possibly in the construction of berry racks) at berry camps. He also noted that the Coast Tsimshian used a decoction of this plant as an emetic and purgative for treating an unspecified type of poisoning. Guédon (1973) has reported that the "cones" of mountain alder were eaten raw as a treatment for vomiting blood.

Identified by Smith as: <u>Betula</u> sp. (Birch)—there is only one kind of birch in the Gitksan country (Luke Fowler)

Smith's transcription of Gitksan name: amhawwak, am meaning "good" and hawwak, the "outer bark"

Modern spelling of Gitksan name: amhaawa<u>k</u> (literally, 'good for-paper birch bark')

Among the Gitksan, according to Luke Fowler, May 24, 1926, birch wood was used for firewood and the bark for pails, large flat basins two feet long, and torches. For a torch, (lāoks [laaxws]),[97] it was rolled (see Figure 20).

<u>Editorial comments</u>: Smith also indicated that birch bark may also have been used folded and used as a spoon.

---

[97]This word means, literally, 'light,' while 'maaxs is the old Gitksan word for a birch bark torch (used for night fishing).

Figure 20. Betula papyrifera (paper birch)

(38) Corylus cornuta Marsh. var. cornuta (Beaked Hazelnut)

Identified by Smith as: (36) Corylus rostrata Ait. (Hazel)

Smith's transcription of Gitksan name: skanjekhh, jekhh meaning basin—the half of a nut shell[98]

Modern spelling of Gitksan name: sgants'ek' (WG), sgants'ak' (EG) (literally, 'hazel nut [or dish] plant')[99]

Among the Gitksan, according to Luke Fowler, May 24, 1926, hazel nuts were gathered and kept all winter to eat raw. The roots were bent to form a hooked stick, guntl [gonhl], used in ground hockey which was played with one round stone by about twenty boys each armed with a stick. The leafed branches were spread like a mat on which to clean salmon. No medicine was made of the hazel (see Figure 21).

---

[98] The word that Smith reported here (i.e., jekhh) seems similar to another term (i.e., jĕuk) which Smith reported to refer to an unidentified plant (see entry #112 and the following footnote).

[99] The term ts'ek' (WG), ts'ak' (EG) means both 'hazel nut' and 'dish.' It is possible that the Gitksan name for hazel may be applied in recognition of the resemblance of the nut shell and a dish. Or, instead, the association of hazel nut and dish in this context may represent a folk etymology where the words for these two items are identical, but etymologically unrelated to one another.

Figure 21. Corylus cornuta (beaked hazelnut)

Brassicaceae (syn.: Cruciferae, Mustard or Crucifer Family)

(39) ?†Sisymbrium officinale (L.) Scop. (Hedge Mustard)
Identified by Smith as: (50) Sysymbrium incisum Eng. (a Hedge Mustard) (too young for exact
    identification; looks like yarrow but no such odour)
Gitksan name: (no name recorded by Smith)

Among the Gitksan, according to Luke Fowler, May 24, 1926, this plant was mashed and put in bad cuts. Luke Fowler did not know its name in Gitksan. It grows on a dry open place where trees are few and where a river has been.

Campanulaceae (Harebell Family)

(40) Campanula rotundifolia L. var. alaskana A. Gray and/or C. rotundifolia L. var. rotundifolia
    (Common Harebell, or Bluebells of Scotland)
Identified by Smith as: (133) Campanula rotundifolia L. (Bluebell, Harebell)
Smith's transcription of Gitksan name: his hab ash huit, his means "resembles" and habash
    means "a white sharp long grass that grows in meadows"
Modern spelling of Gitksan name: hishabasxwit ("it's a pretend/false grass")

Among the Gitksan, according to Bob Robinson, October 3, 1926, this plant was of no use.

Caprifoliaceae (Honeysuckle Family)

(41) Lonicera involucrata (Richards.) Banks ex Spreng. (Black Twinberry, or Bearberry
    Honeysuckle)
Identified by Smith as: Lonicera involucrata Banks. (Black Twin-berry)
Gitksan name: (no name recorded by Smith)[100]

Editorial comments: Smith did not include this species in his Gitksan ethnobotany report but elsewhere, (Smith 1927), he reported that "The Gitksan Indians used an infusion of the inner bark of this plant, the bark itself, or the fresh juice of the berries for sore eyes." Later,

---

[100]Guédon (1973) recorded the term maa'ya gaak (literally, 'berry of raven') for this species.

Smith (1929:63) stated that "Fresh juice of berries used in sore eyes. Inner bark (when berries not available) soaked in water, and the milky solution used in the eyes" (see Figure 22).

Figure 22. Lonicera involucrata (black twinberry)

CAUTION—THE FLOWERS AND FRUITS OF THE FOLLOWING SPECIES ARE EDIBLE, BUT OTHER PARTS OF THE PLANT ARE TOXIC!

(42) Sambucus racemosa L. ssp. pubens (Michx.) House var. arborescens and possibly also S. racemosa L. ssp. pubens (Michx.) House var. leucocarpa (Coastal Red Elder or Elderberry)
Identified by Smith as: (33) Sambucus racemosa L. (Red-fruited Elder)
Smith's transcription of Gitksan name: skan läch· (skan lawchs)
Modern spelling of Gitksan name: sganloots' (plant), loots' (fruit)

Among the Gitksan, according to Luke Fowler, May 24, 1926, the berries of the coastal red elder were eaten and the roots were used for an emetic and purgative. The berries were never eaten raw but were boiled and made into cakes which were dried on racks. For this purpose they were sometimes mixed with blue huckleberries (mīga^{r}an [miigan]) or black huckleberries (mae or mų [maa'y]) which are a little different and they were sometimes mixed with both of these. Cakes of mixed coastal red elder berries and black huckleberries were called mae an läch· (mųanlawts)[101]

The dry berry cakes were kept all winter in small boxes. These cakes were the chief of the best foods of the Gitksan Indians. They were expensive and were used by the important and wealthy people at feasts. They were served with eulachon or salmon oil, or groundhog, black bear, or grizzly bear grease.

For medicine the bark of the roots was scraped off and mixed with water but this mixture was not boiled. The infusion when drunk caused vomiting and purging. Like the infusion of red alder bark, it was a great Gitksan medicine and was used for the same purpose (see Figure 23).[102]

---

[101]This term may be maa'y anloots', where maa'y is the general term for berry or any fruit and anloots' means "place where there are elderberries."
[102]Smith noted that these berries were also eaten with dwarf blueberries (Vaccinium caespitosum). Guédon (1973) reported that the red elder purgative would restore one's appetite and eliminate drowsiness.

Figure 23. Sambucus racemosa (red elderberry)

CAUTION—THE FOLLOWING SPECIES IS POTENTIALLY HAZARDOUS IF EATEN!

(43) Symphoricarpos ?albus (L.) Blake var. albus and possibly also S. albus (L.) Blake var. laevigatus (Fern.) Blake (Common Snowberry)

Identified by Smith as: (2) Symphoricarpos racemosus Michx. (Wax Berry, Snow Berry) (from gravel terrace two miles north of Kitwanga, B.C., August 17, 1925)

Smith's transcription of Gitksan name: skangisgitch

Modern spelling of Gitksan name: sgangisgits or sgangisgits' (literally, 'plant [undefined]')[103]

Among the Gitksan, according to John Fowler, August 17, 1925, the hollow shoots were used for pipe stems by old men. The roots, bark, leaves and berries were of no use to the people although the berries are eaten by grouse (see Figure 24).[104]

---

[103]The undefined element of this plant name is similar to the Nisga'a name for "sparrow," gisgits' (possibly in reference to one or more species of Emberizidae and Passeridae) but no Gitksan "sparrow" term is documented.

[104]"Grouse" here may refer to one or more birds that previously have been identified as types of "grouse": 1) the bird known in Gitksan as maxmeek (?Bonasa umbellus [L.], ruffed grouse—the bird likely to be meant here); 2) the bird called litsxw (?Dendragapus canadensis [L.], spruce grouse); or 3) the bird called bisdaÿ (?Lagopus lagopus [L.], willow ptarmigan).

Figure 24. Symphoricarpos albus (common snowberry)

(44) <u>Viburnum</u> <u>edule</u> (Michx.) Raf. (Highbush Cranberry)

Identified by Smith as: (1A) (1B) <u>Viburnum</u> <u>pauciflorum</u> Raf. (Squashberry, High Bush Cranberry) (from Kitwanga, B.C., August 23, 1925)

Smith's transcription of Gitksan name: skan see tips, si tips

Modern spelling of Gitksan name: sgants'idipxst (plant), ts'idipxst (fruit)

Among the Gitksan, according to John Fowler, August 20, 1925, the berries were eaten raw not made into dry cakes or, until recently boiled for jam and the bark and twigs were boiled for a cough decoction, but the roots were of no use. Abraham Fowler, August 9th, said rope was made of this plant.

The raw berries were put in a box with eulachon grease to keep them for winter. According to Abraham Fowler, August 9, 1925, the berries were kept in eulachon grease, were eaten raw, and made into jam.

The bark and twigs when boiled made a very strong cough decoction which was used for consumptives. The dose being only one cup full in the morning (see Figure 25).

<u>Editorial comments</u>: Luke Fowler, June 15, 1926, stated that the twigs of this plant were strong if twisted like redcedar withes and could be used to tie rafts together. He also (May 24, 1926) said that all parts of the plant should be boiled and the decoction used as a physic for any sickness, or with devil's club as a diuretic for individuals who cannot urinate. Squashberry and devil's club were also useful in treating "rupture" (possibly hernia). Kneeling angelica roots were boiled with this plant for an unspecified medicine. He also indicated that the ripe berries were edible and could be mixed with black berries, "little blueberries" (dwarf blueberries), or rose hips to dry in cakes.

Figure 25. <u>Viburnum</u> <u>edule</u> (highbush cranberry)

Celastraceae (Stafftree Family)

CAUTION—THE FOLLOWING SPECIES IS POTENTIALLY HAZARDOUS IF EATEN!

(45) <u>Pachistima</u> <u>myrsinites</u> (Pursh) Raf. (Falsebox, Mountain or Oregon Boxwood, Mountain-
   box, or Mountain-lover)

Identified by Smith as: (121) (157) <u>Pachystima</u> <u>myrsinites</u> Raf. (False Box) (from Nash, B.C.,
   September 6, 1926)

Smith's transcription of Gitksan name: hĭs skan du mē ĭt sit (his skan dĭmē ithut), his means
   "nearly," skan, "plant," dumēĭt, (dĭmē) a red berry [kinnikinnick] that Indians eat raw
   in eulachon oil; sit (ithut) the end of a word

Modern spelling of Gitksan name: hissgant'imi'ytsit, hissgant'imi'ytxwit ("it's a pretend/false
   kinnikinnick")

Among the Gitksan, according to Bob Robinson, September 5, 1926, and October 3, 1926, this plant was of no use and had no name of its own (see Figure 26).

Figure 26. Pachistima myrsinites (falsebox)

Cornaceae (Dogwood Family)

(46) Cornus canadensis L. (Bunchberry, or Canadian Bunchberry)
Identified by Smith as: (11) Cornus canadensis L. (Bunchberry, Dwarf Dogwood, Dwarf Cornell) (from Kitwanga, B.C., August 23, 1925)
Smith's transcription of Gitksan name: kāpcōip[105]
Modern spelling of Gitksan name: gapk'oyp

Among the Gitksan, according to John Fowler, August 20, 1925, this plant was of no use to the people, although chipmunks[106] eat the berries continually. The roots were not used. The plant was useless for medicine, and there was no myth about it.

Editorial comments: According to Luke Fowler, May 25, 1926, the Babine (Athabaskan) people used a decoction of this plant for a medicine for an unspecified illness. The Gitksan ate the berries with grease or used them with other fruits such as dwarf blueberries to make berry cakes.

(47) Cornus stolonifera Michx. (Red-osier Dogwood)
Identified by Smith as: (8) Cornus stolonifera Michx. (Red-osier Dogwood, Red Willow) (from Kitwanga, B.C., 1925)
Smith's transcription of Gitksan name: sconclatl or scin clătl
Modern spelling of Gitksan name: sganxhlaahl (plant)

Among the Gitksan, according to John Fowler, August 27, and Abraham Fowler, August 19, 1925, this bush was of no use. The bark was not used for smoking and the berries were not eaten although both black and grizzly bears eat them.

Among the Gitksan, according to Luke Fowler, May 24, 1926, only the black bears, grizzly bears, and blue jays[107] eat the berries of this plant. The bark, he said, was not smoked by the Gitksan but he said he had heard that the Indians near Victoria had learned to smoke it from the Hudson's Bay Company. The wood was used for bows for children, but not for bows used in war or hunting. The plant was not used as a diuretic but the withes were

---

[105] Also see Disporum hookeri var. oreganum (#102).
[106] Here, "chipmunk" may refer to Tamias amoenus J.A. Allen (yellow-pine chipmunk). "Chipmunk" is known in Gitksan as jilgeesa or jilgeest.
[107] The blue jay, or Steller's jay (Cyanocitta stelleri [Gmelin]) is known in Gitksan as k'alidakhl.

sometimes twisted to use for tying heavy things or as an undermat upon which salmon were cleaned (see Figure 27).

Figure 27. Cornus stolonifera (red-osier dogwood)

Crassulaceae (Orpine Family)

(48) Sedum ?stenopetalum Pursh (Worm-leaved or Narrow-petaled Stonecrop) or, more likely, S. lanceolatum Torr. (Lance-leaved Stonecrop)[108]

Identified by Smith as: (53) (103) Sedum stenopetalum Pursh (Stonecrop) ([53] from rock outcrop near Woodcock, B.C., May 29, 1926; [103] from Kitwanga, B.C., June 9, 1926)

Smith's transcription of Gitksan name: dŭpï'esgāk

Modern spelling of Gitksan name: t'ipyeeshl gaak (literally, 'raven's stonecrop')[109]

Among the Gitksan, according to Abraham Fowler, May 29, 1926, the stems of this plant were pounded up and put on cuts. But according to Robert A. Sampare, Gus Sampare and Abraham Fowler, June 9, 1926, it was of no use and had no name (see Figure 28).

---

[108] Sedum stenopetalum probably does not occur in the Gitksan territory and this may be an erroneous identification where S. lanceolatum is the more probable species (J. Pojar, pers. comm. 1996). Another species of stonecrop—S. divergens S. Wats. (spreading stonecrop)—has been used as food by various Tsimshianic groups including the Gitksan, who refer to this plant as t'ipyees and who eat its leaves (People of 'Ksan 1980:126).

[109] It is not possible to determine with certainty what the final -gāk of Smith's term might mean, although it seems suggestive of gaak, 'common raven.' The term for stonecrop alone (i.e., spreading stonecrop) would be t'ipyeest.

Figure 28. Sedum lanceolatum (lance-leaved stonecrop)

Elaeagnaceae (Oleaster Family)

(49) <u>Shepherdia</u> <u>canadensis</u> (L.) Nutt. (Soopolallie, Soapberry, or Canadian Buffalo-berry)
Identified by Smith as: (19) <u>Shepherdia</u> <u>canadensis</u> Nutt. (Soopolallie, Soapolallie) (Seen on a
    dry creek bed on the terrace on the south side of Skeena River opposite Kitwanga, B.C.,
    August 20, 1925. This creek bed is called Ăngkăish, ish referring to this plant.)[110]
Smith's transcription of Gitksan name: skan is, skan meaning "plant;" ish
Modern spelling of Gitksan name: sganis (plant), is (fruit)

Among the Gitksan, according to John Fowler, August 20, 1925, the berries were eaten and the roots were used in a medicine for rheumatism but the leaves were of no use.

The berries were eaten raw and were also boiled in a box with hot stones, mashed up, spread on leaves of the thimbleberry, laid like shingles on a rack for drying berry cakes and dried in cakes perhaps a quarter of an inch thick by twelve inches wide by nine and a half feet long. They were boiled half a day and until the next morning. To eat these a few berries or a little of a dry cake of the berry jam was whipped up with plenty of water and a little sugar. This made a froth considered as a delicacy by the Gitksan and neighbouring tribes, especially by the children, and sometimes called Indian ice cream.

For a medicine for rheumatism the roots were boiled entire with twigs of spruce on which were both leaves and bark. A very strong decoction was made. The dose was one cup full three times a day.

Among the Gitksan, according to Bob Robinson, the berries were made into dry cakes for later use, the bark was used for a chronic cough medicine. For this cough medicine the branches with the leaves, that is the whole plant except the roots, were put in a box with water and hot stones were added to cause boiling.[111] Lately the white man's kettle was used over a fire. A large cup full of the decoction was taken three times a day. This stopped the cough, and acted as a physic but not as a diuretic.

---

[110]This place name can be reconstituted as anga'is, "place where there are soapberries." This word is also the noun for "urine" and the verb "urinate." Some Gitksan consultants draw the connection between the foam of urine and the foam produced by whipping soapberries.

[111]Guédon (1973) has also reported that soapberry branches could be used as a laxative or, when combined in a decoction with alder (perhaps red alder) and hybrid Sitka spruce bark, as an arthritis treatment and appetite stimulant.

Among the Gitksan, according to John Fowler, August 20, 1925, the branches and blossoms were of no use. A few years ago the Gitksan learned of wine and made it from the berries but the authorities objected to the practice. The fruit was eaten raw because there was not enough of it to cook or make into dry cakes (see Figure 29).

Editorial comments: Elsewhere, Smith indicated that soapberry spoons were made from the wood of Douglas maple.

Figure 29. Shepherdia canadensis (soapberry)

Ericaceae (Heath Family)

(50) <u>Arctostaphylos</u> <u>uva-ursi</u> (L.) Spreng. (Kinnikinnick, Common Bearberry, Mealberry, or Sandberry)

Identified by Smith as: (3) (14) <u>Arctostaphylos</u> <u>uva-ursi</u> Spreng. (Kinnikinnik, Bear Berry) (from Kitwanga, B.C., August 23, 1925)

Smith's transcription of Gitksan name: scontŭmĕĕt, the berries being called tŭmĕĕt; ɒumēĭt

Modern spelling of Gitksan name: sgant'imi'yt (plant), t'imi'yt (fruit)

Among the Gitksan, according to John Fowler, August 27, 1925, the ripe berries were eaten raw, and were stored for winter in a box containing eulachon oil, but were not boiled. The leaves were mixed with tobacco and smoked by some old men but the roots and bark were of no use nor was any part of the plant made into medicine. Luke Fowler said that the leaves of this plant were not smoked in the early days and that the Hudson Bay introduced this idea to the Gitksan.

The berries kept in eulachon grease were used all winter according to John Laknitz. John Fowler said that this was the only plant that was smoked and before the coming of the white man the Gitksan had no tobacco. John Fowler on August 9th said the ripe berries were eaten raw and kept in eulachon grease and the leaves were smoked (see Figure 30).

Figure 30. Arctostaphylos uva-ursi (kinnikinnick)

(51) Ledum groenlandicum Oeder (Labrador Tea)

Identified by Smith as: (144) Ledum groenlandicum Oeder. (Labrador Tea, Hudson Bay Tea)

(from muskeg around hospital lake near Hazelton, B.C., August 29, 1926)

Smith's transcription of Gitksan name: skan da tōātł, skan means plant.

Modern spelling of Gitksan name: sgandaxdo'ohl (plant)

Among the Gitksan, according to Bob Robinson, September 5, 1926, the leaves of this plant were boiled for a beverage and a diuretic. Formerly they were boiled in a woven basket or wooden box by means of hot stones, but now a white man's kettle is used (see Figure 31).

Figure 31. <u>Ledum</u> <u>groenlandicum</u> (Labrador tea)

(52) <u>Menziesia ferruginea</u> Sm. ssp. <u>ferruginea</u> (False Azalea)
Identified by Smith as: (156) <u>Menziesia ferruginea</u> Sm. (False Azalea)
Smith's transcription of Gitksan name: skan teat^u, skan means "plant," teat^u "thunder"
Modern spelling of Gitksan name: sgantya'ytxw (literally, 'thunder plant')

    Among the Gitksan, according to Bob Robinson, September 5, 1926, this plant was of no use.[112]

(53) <u>Oxycoccus oxycoccus</u> (L.) MacM. (Bog Cranberry)[113]
Identified by Smith as: (22) <u>Vaccinium oxycoccus</u> var. <u>intermedium</u> Gray (Cranberry)
Smith's transcription of Gitksan name: miat (meought); John Fowler said that this plant grows
    in round muskegs which were called lalaho or lalahe [laalax̱'u or laalax̱'i].
Modern spelling of Gitksan name: 'mii'oot

    Among the Gitksan, according to John Fowler, September 16, 1925, the berries were used for food but the roots and bark were of no use. No part of the plant was used for medicine, and it had no myth. The berries were eaten raw and were also boiled in a box with hot stones to make jam which was eaten with eulachon grease. According to Luke Fowler, June 8, 1926, these berries could be eaten raw or with grease (ɬaem miăt [hlayim mii'oot])[114] or any kind, including bear, eulachon, salmon, or groundhog. The berries could be kept all winter, until April, if stored in grease without being boiled first, which was the method used before there were any white men in the area (see Figure 32).

---

[112]The Coast and Southern Tsimshian, however, recognized and ate galls of the fungus <u>Exobasidium</u> sp. affin. <u>vaccinii</u> on <u>M. ferruginea</u> (Compton 1995).

[113]Instead of this species, People of 'Ksan (1980:66, 67) have identified the referent of "'mii oot"as "<u>Vaccinium vitis-idaea</u>, low bush cranberry" (i.e., <u>V. vitis-idaea</u> L. ssp. <u>minus</u> [Lodd.] Hult. [lingonberry, or rock or mountain cranberry]).

[114]This term refers to the dish consisting of bog cranberries mixed with eulachon grease.

Figure 32. Oxycoccus oxycoccus (bog cranberry)

(54) <u>Vaccinium</u> ?<u>alaskaense</u> Howell (Alaskan Blueberry) and/or <u>V</u>. ?<u>ovalifolium</u> Sm. (Oval-leaved Blueberry)[115]

Identified by Smith as: "high bush blueberry" or "blue huckleberries"

Smith's transcription of Gitksan name: skan mī gan, migaʳan

Modern spelling of Gitksan name: sganmiigan (plant), miigan (fruit)[116]

<u>Editorial comments</u>: In comments associated with <u>Sambucus racemosa</u> (#42) Smith reported that "blue huckleberries" were sometimes mixed with coastal red elder berries and black huckleberries (see Figures 33 and 34).

---

[115] Elsewhere, the species referred to as "'mīi gan" has been identified as <u>V</u>. <u>ovalifolium</u> (People of 'Ksan 1980:68, 69). While both <u>V</u>. <u>alaskaense</u> and <u>V</u>. <u>ovalifolium</u> occur within Gitksan territory, the latter species is more common (J. Pojar, pers. comm. 1996).

[116] This plant name includes the element gan, which be itself means '(straight) tree, wood, log, etc.' It is not clear, however, that it has the same meaning here.

Figure 33. <u>Vaccinium</u> <u>alaskaense</u> Howell (Alaskan blueberry)

Figure 34. Vaccinium ovalifolium (oval-leaved blueberry)

(55) <u>Vaccinium</u> <u>caespitosum</u> Michx. (Dwarf Blueberry or Bilberry)

Identified by Smith as: (52) (78) (160) <u>Vaccinium</u> <u>caespitosum</u> Michx. (Dwarf Bilberry) (from near Woodcock, B.C., May 29, 1926)

Smith's transcription of Gitksan name: mīgetɬ, mīetl or mīet

Modern spelling of Gitksan name: 'miyehl (WG), 'miyahl (EG)[117] (fruit)

Among the Gitksan the berries of this plant were eaten raw and in the form of dried cakes. Sometimes they were mixed with berries of coastal red elder, bunchberry, highbush cranberry, or wild cherry (scon snāō), but not with those of choke cherry (hallochalk). The dried cakes were served with oil of eulachon, salmon, groundhog, or bear (see Figure 35).

---

[117]Rigsby has also recorded a variant, miyehl (WG) with a soft initial nasal.

Figure 35. Vaccinium caespitosum (dwarf blueberry)

(56) <u>Vaccinium</u> ?<u>membranaceum</u> Dougl. ex Hook. (Black Huckleberry or Blueberry)
Identified by Smith as: "black huckleberries," "black berries" or "black bilberries"
Smith's transcription of Gitksan name: mae or my
Modern spelling of Gitksan name: maa'y (literally, 'berry,' i.e., any berry or other fruit)

<u>Editorial comments</u>: Smith did not identify this species in botanical Latin terms. However, his comments regarding "black berries," "black bilberries" or "black huckleberries" that appear in association with other species probably refer to <u>V</u>. <u>membranaceum</u>.

Fabaceae (syn.: Leguminosae, Bean or Legume Family)

CAUTION—THE FOLLOWING SPECIES IS POTENTIALLY HAZARDOUS IF EATEN!

(57) <u>Lathyrus</u> <u>ochroleucus</u> Hook. (Creamy or Cream-colored Peavine)
Identified by Smith as: (38) <u>Lathyrus</u> <u>ochroleucus</u> Hook. (Yellow Pea)
Smith's transcription of Gitksan name: haegimgish^u (highgimgish^u) "wiping leaves" (the purple one has the same name and use)[118]
Modern spelling of Gitksan name: hagimgasxw (literally, 'wiper')[119]

Among the Gitksan, according to Luke Fowler, May 24, 1926, this plant was used to wipe salmon or meat, but the peas were not used, the people not knowing if they were good for food.

CAUTION—THE FOLLOWING SPECIES IS POTENTIALLY HAZARDOUS IF EATEN!

(58) <u>Lupinus</u> <u>arcticus</u> S. Wats. (Arctic Lupine)
Identified by Smith as: (55) <u>Lupinus</u> <u>arcticus</u> Wats. (Lupine) (from Hazelton, B.C., June 8, 1926)
Smith's transcription of Gitksan name: kāūch (kowuch) (nas in rule)
Modern spelling of Gitksan name: (unrecognizable)[120]

---

[118] The "purple one" referred to by Smith may be <u>L</u>. <u>nevadensis</u> S. Wats. var. <u>pilosellus</u> (Peck) C.L. Hitchc. (purple peavine) or <u>Vicia</u> <u>americana</u> Muhl. ex Willd. (American vetch), two locally common purple-flowered species. Guédon (1973) also associated <u>L</u>. <u>ochroleucus</u> and <u>L</u>. <u>nevadensis</u> with the Gitksan name reported by Smith and reported that <u>L</u>. <u>ochroleucus</u> was eaten by horses and used to clean slime from fish or dust and dirt from western hemlock bark.
[119] This is derived from the singular transitive verb gimk 'wipe,' and it might be translated as 'wiper.' "Wipe" here has a general sense, and one should not infer that people used peavine leaves to wipe their anuses after defecating, as there is a specific word for that action, viz., gam'in.
[120] The term or terms reported by Smith cannot be reconstituted from Smith's transcription and no comparable Tsimshianic lupine terms have been recorded elsewhere. However, the first term reported by Smith seems suggestive of the Nuxalk term

Among the Gitksan, according to Luke Fowler, June 8, 1926, the whole root was eaten in the spring before the leaves came out and in recent years the beans were eaten. A fire was made, stones were heated. These were put in a hole in the ground covered with grass. A little water was put on the stones, so they would steam and the roots were put on the grass. They were covered with a mat woven of redcedar bark and earth, and left over night to bake. In serving these cooked roots oil of eulachon, salmon groundhog, or bear was used. Chinese men taught the Gitksan to boil the beans for food (see Figure 36).

---

q'akwts reported by Nater (1990) for "blue lupine" and identified by Turner (1973) as Lupinus nootkatensis Donn. var. fruiticosus Sims (blue lupine) flowers.

Figure 36. Lupinus arcticus (arctic lupine)

Geraniaceae (Geranium Family)

(59) <u>Geranium richardsonii</u> Fisch. & Trautv. (White or Richardson's Geranium or Crane's-bill)
Identified by Smith as: (126) <u>Geranium Richardsonii</u> F. & M. (White Geranium) (from Kitwanga, B.C., 1926)
Smith's transcription of Gitksan name: his mĭskätuit according to Bob Robinson. His means nearly and mĭskätuit is a plant, the leaves of which are used for drying berries, probably 170 thimbleberry
Modern spelling of Gitksan name: hisnisk'o'otxwit ("it's a pretend/false-thimbleberry")[121]

Among the Gitksan, according to Gus Sampare, September 30, 1926, and Bob Robinson, October 3, 1926, this plant had no use and no name. Bob Robinson said it grows in meadows and swamps.

Grossulariaceae (Currant or Gooseberry Family)

(60) <u>Ribes</u> ?<u>bracteosum</u> Dougl. ex Hook. (Stink Currant) or, more likely, <u>R. laxiflorum</u> Pursh (Trailing Black Currant)[122]
Identified by Smith as: <u>Ribes bracteosum</u>
Smith's transcription of Gitksan name: skan win gilis; the Kuldi [Kuldo] people call it skan similaw[123]
Modern spelling of Gitksan name: (unrecognizable)[124]

<u>Editorial comments</u>: Some of the data presented by Smith for stink current may actually apply to the spiny <u>Ribes</u> <u>lacustre</u> (Pers.) Poir. in Lamarck (black or black swamp gooseberry) rather than to the unarmed species, <u>R</u>. <u>bracteosum</u> and/or <u>R</u>. <u>laxiflorum</u>. Luke Fowler said that the plant known as skan win gilis was a gooseberry or a near relative, has many "prickers that make bad sores," and "has black berries and we do not eat them" because they are "no good." He also stated that he had seen many people boil this plant for a medicine but that he did not know which sickness it was used for. Elsewhere, Smith (1929:58) reported the following

---

[121] This is probably a nonce form.

[122] <u>Ribes bracteosum</u> is a coastal species that occurs in some isolated local populations within Gitksan territory, such as at Kuldo (J. Pojar, pers. comm. 1996). Otherwise, it would seem more likely that <u>R</u>. <u>laxiflorum</u> is the species that would have been recognized and used by the majority of Gitksan people.

[123] This is the only explicit indication Smith gave of dialectal variation in the Gitksan botanical terms he recorded.

[124] These terms are unrecognizable, but the construction of win plus a noun means "place where there is X." The latter term which Smith attributed to the Kuldo people might be Athapaskan in origin but its provenience is unclear.

medicinal information for black gooseberry: "Bark boiled and used as a remedy for some unspecified malady."

(61) <u>Ribes hudsonianum</u> Richards. in Frank. (Northern Blackcurrant)
Identified by Smith as: (41) (87) <u>Ribes Hudsonianum</u> Richards (Black Currant)
Smith's transcription of Gitksan name: mīishwut, meaning pungent berries. The bark and berries smell and taste pungent.[125]
Modern spelling of Gitksan name: mii'isxwit (literally, 'stinking berries')

Among the Gitksan, according to Luke Fowler, May 24, 1926, the berries of this plant were eaten raw and sometimes were made into thin cakes and dried. The dry cakes were served with any kind of oil.

According to Bob Robinson, July 24, 1926, the currants were eaten raw when one was hunting, but were never boiled or cooked in any way. No medicine was made of the plant and the leaves and roots were of no use.

(62) ?<u>Ribes lacustre</u> (Pers.) Poir. in Lamarck (Black or Black Swamp Gooseberry) or ?<u>Ribes oxyacanthoides</u> L. ssp. <u>cognatum</u> Greene and/or <u>R</u>. <u>oxyacanthoides</u> ssp. <u>oxyacanthoides</u> (Northern or Northern Smooth Gooseberry)[126]
Identified by Smith as: (138) <u>Ribes oxyacanthoides</u> L. (Smooth Gooseberry)
Smith's transcription of Gitksan name: skandĭllūsha (Luke Fowler) skan da losa (Bob Robinson); skan means plant (Bob Robinson) the berries being called dĭllūsha (Luke Fowler) dalosa (Bob Robinson)
Modern spelling of Gitksan name: sgandilusa'a (plant), dilusa'a, dilusa (fruit)

Among the Gitksan, according to Luke Fowler, May 24, 1926, there is only one kind of gooseberry[127] and it was used the same as red raspberries, eaten raw as well as made into cakes and dried. Sometimes black bilberries were mixed with them when making cakes. Before

---

[125]Here, Smith also reported that "There is another kind, which grows on the mountains, called jĕuk and another which grows low down called migānao [miiganaa'w] meaning frog berries skan mī nō' (skan meknow' [sganmiiganaa'w]), skan meaning plant. There are two kinds of plant of this name. The other has large leaves. They also have a pungent odour." (Also see unidentified plant, #112). Smith's skan mī nō' may be the otherwise unknown and undefined form sganmiino'o or it could perhaps represent an erroneously written version of the "frog berries" term.

[126]The People of 'Ksan (1980:74, 75) have identified the referent of the Gitksan terms shown below as "wild gooseberries," <u>Ribes lacustre</u>.

[127]Despite this comment, Smith also reported that another species of <u>Ribes</u>—<u>R</u>. <u>hudsonianum</u> (northern blackcurrant)—was used as food.

they were eaten the cakes were soaked in water and then served with oil of eulachon, salmon, groundhog, or bear.

According to Bob Robinson, October 3, 1926, gooseberries were considered a nice sweet fruit and were eaten raw but were not made into dry cakes, or kept for winter use. The raw gooseberries were served in eulachon oil.

Lamiaceae (syn.: Labiatae, Mint Family)

(63) <u>Mentha arvensis</u> L. (Field Mint)
Identified by Smith as: (64) (148) (159) <u>Mentha canadensis</u> L. var. <u>glabrata</u> Benth. (Canada Mint)
Smith's transcription of Gitksan name: skan ēsut, skan means plant; ēsut (īsxut C.M. Barbeau)[128] stink; two women say just mezeroolē [i.e., "flower"]
Modern spelling of Gitksan name: sgan'isxwit (literally, 'stinking plant'), majagalee (literally, 'flower')

Among the Gitksan, according to Bob Robinson, September 5, 1926 this plant was of no use. Luke Fowler, June 8, 1926 and two women said the same, and also that it had no name other than mezerulē, flowers.[129]

(64) <u>Prunella vulgaris</u> L. ssp. <u>lanceolata</u> (Bart.) Hult. (Self-heal)
Identified by Smith as: (82) <u>Prunella vulgaris</u> L. (Heal-all)
Smith's transcription of Gitksan name: his cowat sut, his meaning resembling, cowat, Indian carrots; Indian carrots grow close to Skeena River and up on the mountains and are not up by July 24th.
Modern spelling of Gitksan name: hisk'awtsxwit ("it's a pretend/false carrot")[130]

Among the Gitksan, according to Bob Robinson, July 24, 1926, this plant was of no use.

---

[128]C.M. Barbeau was an ethnologist with the Division of Anthropology of the Geological Survey of Canada. Smith's mention of his name here and in association with a story regarding <u>Geocaulon lividum</u> (#94) suggests that these two men conversed about their various projects in the area—Smith's on ethnobotany, archaeology and totem pole restoration, and Barbeau's on totem poles. Apparently, Barbeau must have offered at least minimal assistance regarding Smith's transcription of Gitksan terms.
[129]Other plants referred to simply as "flower" or "flower plant" include <u>Corydalis aurea</u> ssp. <u>aurea</u>, <u>Parnassia palustris</u>, <u>Clematis occidentalis</u> ssp. <u>grosseserrata</u> and <u>Heuchera glabra</u> (see Appendix 4, #30, #34, #35, #41, #45).
[130]The Gitksan word for "carrot" (i.e., <u>Daucus carota</u> L., domesticated carrot) is k'awts, and its origin is unclear.

Nymphaeaceae (Water-lily Family)

(65) <u>Nuphar polysepalum</u> Engelm. (Rocky Mountain Cow-lily, Spatterdock, Yellow Pond-lily)
Identified by Smith as: ?<u>Nymphaea polysepala</u> (Engelm.) Greene ("Water Lily")
Smith's transcription of Gitksan name: kăɬtăch
Modern spelling of Gitksan name: gahldaats

Among the Gitksan, according to Luke Fowler, May 25, 1926, scrapings of the toasted "root" of the water lily were put in water and the infusion was drunk for hemorrhage of the lungs, and was taken by men to prevent conception (see Figure 37).[131]

---

[131] Elsewhere, Smith (1929:56) reported the following additional medicinal information for yellow pond-lily: "Infusion of scrapings of toasted root (or according to another informant, heart of root, boiled) taken internally for hemorrhage of the lungs and as a contraceptive." Guédon (1973) has reported that in a decoction of yellow pond lily rhizomes was used as a medicine for hemorrhage from tuberculosis. In addition, yellow pond lily rhizomes were applied to treat broken bones, sometimes in conjunction with hybrid Sitka spruce pitch and cow-parsnip. The crushed rhizomes were applied directly in the treatment of arthritis.

Figure 37. Nuphar polysepalum (yellow pond-lily)

Onagraceae (Evening-primrose Family)

(66) Epilobium angustifolium L. ssp. angustifolium and/or E. angustifolium L. ssp. circumvagum Mosquin (Fireweed)

Identified by Smith as: (24) Epilobium angustifolium L. (Fire-Weed) (Seen on the sides of the valley near Kitwanga, B.C., September 16, 1925.)

Smith's transcription of Gitksan name: hasht

Modern spelling of Gitksan name: haast

Among the Gitksan, according to John Fowler, September 16, 1925, in May the inside only of the young shoots was eaten raw.

According to Robert A. Sampare the fibre was used for making cord for fish nets but it was poor material for that purpose. John said the plant had no myth. It is generally known, however, that it is used as a name of a phratry and is shown as a crest on one of the totem poles of Kitseguecla.

Editorial comments: Smith also recorded that "They eat the inside raw scraped with the shell of a mussel[132] from the sea in June until middle of July when it begins to get too old." The part that was left was then put out to dry. It was gathered during the last part of June and in July for use in making thread. According to Bob Robinson, July 18, 1926, the thread was used for making nets and in combination with goat or sheep[133] wool for the long element for making thread used in pack straps (tră dektɬ [tk'aadakhl]). Luke Fowler (June 15, 1926) stated that the cord made from this plant may consist of six or more strands braided and used for tying blankets and boxes. He also said that one must collect the plant for thread before it flowers because after it blooms it becomes too hard to use.

Polygonaceae (Buckwheat Family)

(67) †Rumex acetosella L. (Sheep Sorrel)

Identified by Smith as: (73) Rumex Acetosella L. (Common Sorrel)

Smith's transcription of Gitksan name: gläwkătch, clawkach, "small rhubarb"[134]

---

[132] This mussel is likely Mytilus edulis L. (edible or blue mussel), known in Gitksan as gals.

[133] Here "sheep" may refer to some species of Ovis while actually the mountain goat (Oreamnos americanus [Blainville]) is probably meant.

[134] The "rhubarb" referred to here is likely Rheum rhabarbarum while the actual referent of the Gitksan term shown here is likely Rumex occidentalis S. Wats., western dock. Comparable Tsimshianic and Wakashan terms refer to a variety of referents,

Modern spelling of Gitksan name: tl'ok̲'ats

Among the Gitksan, according to Luke Fowler, June 15, 1926, this plant was boiled, mashed and eaten mixed with oil of eulachon, salmon, groundhog, or bear. It was also boiled, mashed and spread in thin cakes on leaves of the thimbleberry laid overlapping on wooden racks and then dried in the sun over the smoke of a slow fire. Previously, May 24, he has stated that it was not kept for winter.

Ranunculaceae (Buttercup Family)

CAUTION—THE FOLLOWING SPECIES IS TOXIC!

(68) Actaea rubra (Ait.) Willd. (Baneberry)
Identified by Smith as: (75) (110) Actaea arguta Nutt. (Baneberry) (from Hazelton, B.C., July 18, 1926)
Smith's transcription of Gitksan name: skan mae ya smah, skan meaning plant, mae meaning berry, and smah meaning black bear (Bob Robinson)
Modern spelling of Gitksan name: sganmaa'ya smex (WG), sganmaa'ya smax (EG) (literally, 'black bear berry plant')

Among the Gitksan, according to Luke Fowler, June 8, 1926, and Bob Robinson, July 18, 1926, this plant was of no use. Bob Robinson said the berries are eaten by black bears. Luke Fowler said it had no name.

CAUTION—THE FOLLOWING SPECIES IS POTENTIALLY HAZARDOUS IF EATEN!

(69) Anemone multifida Poir. (Cut-leaved or Pacific Anemone)
Identified by Smith as: (52A) (112) Anemone globosa Nutt. (A. multifida Poir.) (Wind-flower) ([112] From near Hazelton, B.C., May 24, 1926, [52A] From near Woodcock, B.C., May 29, 1926)
Smith's transcription of Gitksan name: lam
Modern spelling of Gitksan name: laam[135]

---

including tobacco, garden rhubarb and western dock, the latter having been used as a wild source of edible greens (Compton 1993; Dunn 1978).

[135]This is a loan word from Chinook Jargon, and its Jargon source was English "rum."

Among the Gitksan, according to Luke Fowler, May 24, 1926, handsful of this plant were eaten in the sweat bath when that was employed for curing rheumatism. A decoction of the plant was sometimes used for the same purpose. It burns the tongue and consequently when these people first tasted whisky they said it tasted like lam and gave it the same name. According to Abraham Fowler, May 29, 1926, this plant was of no use and had no name.

(70) Aquilegia formosa Fisch. in DC. (Red or Sitka Columbine)
Identified by Smith as: (40) Aquilegia formosa Fischer (Columbine)
Smith's transcription of Gitksan name: ik lē ām chŭk, (ik lay am chuck), "good for bleeding nose"
Modern spelling of Gitksan name: ihlee'em ts'ak̲ (literally, 'bleeding nose')[136]

Among the Gitksan, according to Luke Fowler, May 24, 1926, the nectar in the little round knobs on the end of the flower spurs was eaten for a sweet, like candy.

CAUTION—THE FOLLOWING SPECIES IS TOXIC!

(71) Delphinium glaucum S. Wats. (Tall Larkspur)
Identified by Smith as: (79) (151) Delphinium Brownii Rydb. (Larkspur) (from Hazelton, B.C., July 18, 1926)
Smith's transcription of Gitksan name: hammok ganao; cla hamok ganao, hamok meaning rhubarb, and ganao, frog
Modern spelling of Gitksan name: ha'mookhl g̲anaa'w ("frog cow-parsnip," literally, 'sucker of frog,' or 'frog-sucker')

Among the Gitksan, according to Bob Robinson, July 24, 1926, and September 5, 1926, this plant is eaten by horses,[137] but was of no use and had no story about it. On September 5, he called it hamok ganao and translated hamok as parsnip.

CAUTION—THE FOLLOWING SPECIES IS TOXIC!

(72) Ranunculus abortivus L. (Kidney-leaved Buttercup)

---

[136]It is unclear as to whether this many be a loan-translation from English "bloody nose." This is the same term as given for Castilleja miniata.
[137]The domesticated horse (Equus caballus L.) is known in Gitksan as gyuwadan.

Identified by Smith as: (45) Ranunculus abortivus L. (a Buttercup) (from near Hazelton, B.C., 1926)

Smith's transcription of Gitksan name: belanwats^u, land otter belt, belan meaning belt, wats^u meaning land otter.

Modern spelling of Gitksan name: bilena 'wats̲x (WG), bilana 'wats̲x (EG) (literally, 'belt of river otter,' or 'river otter-belt')

The seeds of this plant stick to one's clothing. It is eaten by animals. Among the Gitksan, according to Luke Fowler, May 24, 1926, this plant was of no use, not even a medicine being made of it.

(73) Thalictrum occidentale A. Gray (Western Meadowrue)

Identified by Smith as: (39) Thalictrum occidentale Gray (Meadow Rue)

Smith's transcription of Gitksan name: hămōkganao, (hamoak ganao) frog parsnip, ganao meaning frog

Modern spelling of Gitksan name: ha'mookhl ganaa'w ("frog cow-parsnip," literally, 'sucker of frog,' or 'frog-sucker')

Among the Gitksan, according to Luke Fowler, May 24, 1926, a few people used the root for medicine. The leaves or fruit were not used. A small piece of the root was chewed and a little of the juice was swallowed for headache, eye trouble, and sore legs. It cleaned the throat so one could expectorate and possibly accelerated the circulation of the blood.

Rosaceae (Rose Family)

(74) Agrimonia striata A. Michx. (Grooved Agrimony)

Identified by Smith as: (99) (135) Agrimonia gryposepala Wallr. (Agrimony) ([99] from Kitwanga, B.C. June 21, 1926, [135] from near Hazelton, B.C., October 3, 1926)

Smith's transcription of Gitksan name: his hash hoot, nearly fireweed; his means nearly and hasht, 212 fireweed (Bob Robinson)

Modern spelling of Gitksan name: hishaasxwit ("it's a pretend/false-fireweed")[138]

Among the Gitksan, according to Abraham Fowler, June 21, 1926, and Bob Robinson, October 3, 1926, this plant was of no use.

---

[138]This is probably a nonce form.

(75) <u>Amelanchier</u> <u>alnifolia</u> (Nutt.) Nutt. (Saskatoon)

Identified by Smith as: (10) (183) <u>Amelanchier</u> <u>florida</u> Lindl. (Juneberry, Saskatoon) (from Kitwanga, B.C., August 23, 1925)

Smith's transcription of Gitksan name: scongĕm, gĕm meaning berry

Modern spelling of Gitksan name: sgangem (WG), sgangam (EG) (plant), gem (WG), gam (EG) (fruit)

Among the Gitksan, according to John Fowler, August 27, 1925, the berries were eaten, the wood was used for adze handles and the branches for arrows, but the roots and bark were of no use and the plant was useless for medicine. Arrows made of the branches of this bush were especially good, the wood being very strong. They were strong enough to be used by the Indians of long ago for shooting grizzly bears where the rifle is used now.

The berries were eaten raw, or boiled, and were boiled and made into dry cakes. Among the Gitksan, according to Luke Fowler, May 24, 1926, saskatoon berries were eaten raw or were made into cakes and dried, the dry cakes being served with eulachon or salmon oil, or groundhog, black bear, or grizzly bear grease. The wood was the main one used for arrows but was not used for adze handles.[139] No medicine was made from this plant (see Figure 38).

---

[139] Elsewhere Smith (1927) stated that "The branches were the chief wood used for arrows as they are strong. Large branches were employed for the handles of spears used in war or for spearing bears."

Figure 38. Amelanchier alnifolia (saskatoon)

(76) _Aruncus dioicus_ (Walt.) Fern. (Goatsbeard)

Identified by Smith as: (80) (162) _Aruncus sylvester_ Kost. (Goat's Beard) ([80] from Hazelton, B.C., July 24, 1926, [162] from Kitwanga, B.C., June 25, 1926)

Smith's transcription of Gitksan name: hĭslēuqŏt, (hĭslayuqut), his meaning similar, and lēok 242 spreading dogbane the plant used for fibre. The name is the same as was given for 219 (84), sweet cicely.

Modern spelling of Gitksan name: his-(unrecognizable, perhaps spreading dogbane)-xwit ("it's a pretend/false-[undefined]")[140]

Among the Gitksan, according to Bob Robinson, July 24, 1926, and Abraham Fowler, June 25, 1926, this plant was of no use, no medicine was made of it and there was no story about it. Abraham said it had no name.

(77) _Crataegus douglasii_ Lindl. (Black Hawthorn)

Identified by Smith as: (16) (182) _Crataegus brevispina_ (Dougl.) Hell. (Black Hawthorn, Haw) (from bottom lands, Kitwanga, B.C., September 3, 1925)

Smith's transcription of Gitksan name: skansnăx, skan snăx or scon snax, snăx meaning the fruit, skan snăx = the tree

Modern spelling of Gitksan name: sgansna<u>x</u> (plant), sna<u>x</u> (fruit)

Smith's transcription of Gitksan term: thorns = (ben ah kt) Ben a kt

Modern spelling of Gitksan term: (unrecognizable, possibly bina<u>k</u>t or binaa<u>k</u>t, "thorns"?)

Among the Gitksan, according to John Fowler, September 3rd, 1925, the apples were eaten, and the wood was used for adze handles, but the roots, bark, leaves and thorns were useless. According to Luke Fowler, May 24, 1926, this plant (possibly the branch with the thorn attached) was used for fishing hooks for trout[141] and the wood has recently been used for axe handles. Only this wood and that of red alder were used for adze handles.

The thorn apples were eaten raw and were also boiled in a box by means of hot stones and kept for winter use. They were not mixed with other foods to preserve them and were not made into dry cakes on a drying frame. According to Luke Fowler, June 15, 1926, the fruits were served with grease (salmon, eulachon, black bear, grizzly bear, or groundhog). If the

---

[140]This is probably a nonce form and is the same term as recorded for _Osmorhiza chilensis_.

[141]Here, "trout" may refer to one or more types of fish, including _Salmo gairdneri_ Richardson (rainbow and steelhead trout), known in Gitksan as milit; _Salvelinus malma_ (Walbaum) (Dolly Varden), known in Gitksan as saabaya'a; and ?_Salvelinus namaycush_ (Walbaum) (lake trout), known in Gitksan as laaxw.

fruits are eaten raw, they will make one sick. They should always be boiled for a long time in a box by adding hot stones, after which time they were smashed and put in a box like mince meat (not with oil) for winter use.

(78) <u>Dryas drummondii</u> Richards. in Hook. var. <u>drummondii</u> and/or <u>D</u>. <u>drummondii</u> Richards. in Hook. var. <u>tomentosa</u> (Farr) Williams (Yellow Mountain-avens)

Identified by Smith as: (119) <u>Dryas Drummondii</u> Rich. (Alpine Avens) (from Nash, B.C., September 6, 1926)

Smith's transcription of Gitksan name: his howōkut, resembling birch, his meaning resembling and howō, birch

Modern spelling of Gitksan name: hishaawak̲xwit ("it's a pretend/false-paper birch")[142]

Among the Gitksan, according to Bob Robinson, October 3, 1926, this plant was of no use.

(79) <u>Fragaria virginiana</u> Duch. ssp. <u>glauca</u> (S. Wats.) Staudt and/or <u>F</u>. <u>virginiana</u> Duch. ssp. <u>platypetala</u> (Rydb.) Staudt (Wild Strawberry)

Identified by Smith as: (32) (122) <u>Fragaria glauca</u> (Wats.) Rydb. (Strawberry) (from Nash, B.C., September 6, 1926)

Smith's transcription of Gitksan name: skan mē gŭn, skan mī gŭnt or skan mīgunt, skan means plant; mī gunt

Modern spelling of Gitksan name: sganmiigwint (plant), miigwint (fruit)

Among the Gitksan, according to Luke Fowler, May 24, 1926, there was only one kind of strawberry, as the frog-berry (<u>Rubus pubescens</u>) is not a strawberry, but is only a little different from it.

Among the Gitksan, according to Bob Robinson, October 3, 1926, strawberries were eaten raw and boiled but the people had no sugar in the early days. The berries were also boiled and made into cakes on rolled leaves of the thimbleberry. These cakes were dried on racks in the sun or over a slow fire. The dried berries were served with eulachon but not with salmon oil or bear grease.

No medicine was made of this plant. The roots were of no use (see Figure 39).

---

[142]This is probably a nonce form.

Figure 39. Fragaria virginiana (wild strawberry)

(80) <u>Malus</u> <u>fusca</u> (Raf.) Schneid. (Pacific Crab Apple)

Identified by Smith as: (17) <u>Pyrus</u> <u>diversifolia</u> Bong. (Wild) (Crab Apple) (from bottom lands, Kitwanga, B.C., September 3, 1925)

Smith's transcription of Gitksan name: skan mīlkst, mīlkst meaning the fruit

Modern spelling of Gitksan name: sganmilkst (plant), milkst (fruit)

Among the Gitksan, according to John Fowler, September 3, 1925, the wood was used only for adze handles for which it is very strong when dry, but the roots, bark and leaves were useless.

The apples were sometimes eaten raw, but they were also boiled in a box with hot stones (always with eulachon grease, (limb mīlkst or łaem mīlkst [hlayim milikst])[143] and kept in a box (see Figure 40).

<u>Editorial comments</u>: Smith also noted that axe handles were formerly made from crab apple wood. Elsewhere, Smith (1929:60) reported the following medicinal information for Pacific crab apple: "Juice, scraped from peeled trunk, used as an eye medicine. Trunk and branches, or scrapings from inside of bark, boiled until thick, and the decoction taken internally over a period of from four to six months for consumption and rheumatism. Said to be a fattening medicine, both laxative and diuretic."

---

[143]This term refers to the dish consisting of crab apples mixed with eulachon grease.

Figure 40. Malus fusca (Pacific crab apple)

(81) <u>Prunus</u> ?<u>emarginata</u> (Dougl.) Walp.(Bitter Cherry) or, more likely, <u>P</u>. <u>pensylvanica</u> L. (Pin or Bird Cherry)[144]

Identified by Smith as: (12) (101) <u>Prunus</u> <u>emarginata</u> Dougl. (Wild Cherry) (from Kitwanga, B.C., August 23, 1925)

Smith's transcription of Gitksan name: snaō; sconsnāō, the cherries being called snāō

Modern spelling of Gitksan name: sgansnaw (plant), snaw (fruit)[145]

Among the Gitksan, according to John Fowler, August 27, 1925, the cherries were considered good to eat raw and were boiled for use as a nice jam but were not made into dry cakes because they were not plentiful enough. Dwarf blueberries were mixed with these fruits but not with those of choke cherry. On the other hand the roots, bark and leaves were of no use, no part of the plant was used for medicine, and it figured in no myth (see Figure 41).

---

[144]This latter species is the most likely one to have been used as food as described by Smith. Bitter cherry fruits are typically too bitter to eat and were not widely used as food among First Nations of British Columbia. In fact, <u>P</u>. <u>emarginata</u> may not occur within Gitksan territory where <u>P</u>. <u>pensylvanica</u> is the predominant wild cherry species (J. Pojar, pers. comm. 1996).

[145]The Western Gitksan forms have not been attested, but they would probably be sgansnew and snew.

Figure 41. Prunus pensylvanica (pin cherry)

(82) <u>Prunus</u> <u>virginiana</u> L. ssp. <u>demissa</u> Taylor & MacBryde and/or <u>P</u>. <u>virginiana</u> L. ssp.
    <u>melanocarpa</u> (Nels.) Taylor & MacBryde (Choke Cherry)
Identified by Smith as: (1) <u>Prunus</u> <u>demissa</u> Nutt. (Choke Cherry) (from gravel terrace two
    miles north of Kitwanga, B.C.)
Smith's transcription of Gitksan name: hăllo chalk (jock), halo chalk or hallochalk
Modern spelling of Gitksan name: haluuts'oo<u>k</u>'[146]

Among the Gitksan, according to John Fowler, August 17, 1925, the cherries were eaten, the cherry pits used for shot, and the wood for adze handles but the roots, bark, and leaves were of no use and medicine was not made from this plant.

The cherries were eaten raw by boys but not by men or women. The cherry pits were used by boys for shot in blowguns made of the hollow stalks of the cow-parsnip, which was called hămäwk. The wood of the larger plants was used for adze handles because it is strong. It was also used for firewood (see Figure 42).

---

[146]The form shown here possibly may be derived from the intransitive verb stem ts'oo<u>k</u>', "be stained" or "be sticky." Its initial prefix is ha-, which derives instrumental nominals, while the proclitic luu means the action or state signified by the stem occurs "inside" something. The more commonly recorded name for choke cherry is miits'oo<u>k</u>'.

Figure 42. Prunus virginiana (choke cherry)

(83) Rosa acicularis Lindl. ssp. sayi (Schwein.) W.H. Lewis (Prickly Rose) and/or R. nutkana
Presl var. nutkana and/or R. nutkana Presl var. hispida Fern. (Nootka Rose)
Identified by Smith as: (23) Rosa sp. (Rose) (There is only one kind near Hazelton, although the blossoms of some are almost white [Luke Fowler].) (Seen on gravel terrace near Kitwanga, B.C., September 16, 1925.)
Name: Skankālums, Kālums meaning the rose apples[147]
Modern spelling of Gitksan name: sgank'alamst (plant), k'alamst (fruit)

Among the Gitksan, according to Luke Fowler, May 24, 1926, rose fruits were eaten raw and were also boiled and made into dried cakes the same as lawts, coastal red elder berries. Squashberries or black berries were mixed with them for this purpose. The wood, being hard and light, after it was dried was used for arrow points for shooting bears and men. The arrow was pulled out of the wound leaving the point in it. Arrows with such points would go straight because the points were not heavy. Stone was never used for arrow points. The rose wood arrow-points were polished with stalks of the horsetail.

Among the Gitksan, according to John Fowler, September 16, 1925, the rose apples were eaten but the roots and bark were of no use, no part of the plant was used for medicine and it figures in no myth.

The rose apples were eaten in the same way as cranberries that is raw and also boiled in a box with hot stones to make jam which was served with eulachon grease.

(84) Rubus idaeus L. ssp. melanolasius (Dieck) Focke (Red Raspberry, or American Red Raspberry)
Identified by Smith as: (68) Rubus idaeus (Red Raspberry) (Seen at Hazelton, B.C., 1926.)
Smith's transcription of Gitksan name: hkan năsĭk (nasick), the fruit being called năsĭk. (Luke Fowler called it strawberry)[148]
Modern spelling of Gitksan name: sgannaasik' (plant), naasik' (fruit)

Among the Gitksan, according to Luke Fowler, May 24, 1926, this plant was used in the same way as the gooseberry, that is, the berries were eaten raw and were made into cakes and

---

[147] The fruiting structures of roses are typically referred to as hips, rather than apples.

[148] Here the red raspberry term is erroneously associated with strawberry (see Fragaria virginiana, wild strawberry).

dried. Sometimes black bilberries were mixed with them when making cakes. For eating the cakes were soaked in water and served with oil of eulachon, salmon, groundhog, or bear.

Among the Gitksan, according to Luke Fowler, June 8, 1926, the berries were eaten raw and although not boiled were mashed up and made into cakes by spreading them on a bed of overlapping thimbleberry leaves laid on a rack and dried over a smoking fire, preferably in the sun.

The dry cakes were served by being broken up in water and mixed with oil of eulachon, salmon, bear, or groundhog (see Figure 43).

Figure 43. Rubus idaeus (red raspberry)

(85) Rubus parviflorus Nutt. (Thimbleberry)

Identified by Smith as: (21) Rubus parviflorus Nutt. (Thimbleberry, Red Cap) (Seen on the gravel terraces near Kitwanga, B.C. August 20, 1925.)

Smith's transcription of Gitksan name: nīgājŭ (kneegaja)

Smith's transcription of Gitksan name: skan nez gä (skan nez gaw^k), leaves = nez gä, berries = gä

Smith's transcription of Gitksan name: skan nizgā, nisqā being the name of the leaves and gä that of the berries

Modern spelling of Gitksan name: sgannisk'o'o (plant), nisk'o'o (fruit)

Among the Gitksan, according to John Fowler, August 20, 1925, the fruit was usually eaten raw because there was not enough of it to cook or make into dry cakes. When dried, the berries were served with eulachon oil but not with salmon oil or bear grease. The leaves were used for shingling on the frames used for drying berry cakes, especially soopolallie berries but the branches, blossoms, and roots were of no use. A few years ago the Gitksan learned of wine and made it from the berries but the authorities objected to the practice.

Among the Gitksan, according to Luke Fowler, June 15, 1926, the leaves of the thimbleberry were tied in a ball, with twine made of willow bark, for men, or boys or girls to use in playing. The leaves were also folded, bitten and unfolded to make designs.

(86) Rubus pubescens Raf. (Trailing Raspberry, or Dwarf Red Blackberry) or possibly R. pedatus Smith (Five-leaved Creeping Raspberry)

Identified by Smith as: (42) Rubus pubescens Raf. (A raspberry, Dwarf Salmonberry) (from Hazelton, B.C.)

Smith's transcription of Gitksan name: miĭt^{u}ganao, frog berries, miĭt^u meaning berries and ganao meaning frog

Modern spelling of Gitksan name: maa'ytxwhl ganaa'w (literally, 'the berries frogs gather')

Smith's transcription of Gitksan name: skan mī nō' (skan meknow'), skan meaning plant; mīgānao, frog berries

Modern spelling of Gitksan name: sganmiiganaa'w (plant), miiganaa'w (fruit, literally, 'berry of frog,' or 'frog-berry')[149]

---

[149]This term could also be written as maa'ytxwhl ganaa'w.

Among the Gitksan, according to Luke Fowler, May 24, 1926, the few berries found are eaten raw never cooked. This bush grows in wet woods. The berries are like salmonberries. When they are ripe there are only a few of them and some years none.

(87) <u>Rubus</u> <u>spectabilis</u> Pursh (Salmonberry)
Identified by Smith as: (67) <u>Rubus</u> <u>spectabilis</u> Pursh. (Salmonberry)
Smith's transcription of Gitksan name: mĭg^räsht
Modern spelling of Gitksan name: miik'ooxst (literally, 'berry maple'?)[150]

Among the Gitksan, according to Luke Fowler, June 8, 1926, salmonberries were eaten raw but were not made into dry cakes.

Among the Gitksan, according to Luke Fowler, May 24, 1926 there was only one kind of salmonberry.[151] The berries of this plant were eaten raw and although they were not cooked they were formerly made into cakes and dried for winter use. No medicine was made of this plant (see Figure 44).

---

[150] While k'ooxst by itself means 'Douglas maple,' it is unclear that it has the same meaning here.
[151] Here Smith may have been referring to the fact that only a single Gitksan term seems to exist for salmonberries. Because of a fruit colour polymorphism, salmonberry produces fruits of different colours. Several First Nations of the Pacific Northwest Coast apply different names to the different coloured fruits.

Figure 44. Rubus spectabilis (salmonberry)

(88) <u>Sorbus</u> <u>sitchensis</u> Roemer (Sitka Mountain-ash) and/or <u>S</u>. <u>scopulina</u> var. <u>cascadensis</u> (G.N. Jones) C.L. Hitchc. Greene (Western Mountain-ash)

Identified by Smith as: (25) <u>Pyrus</u> <u>sitchensis</u> (Roem) Piper or <u>P</u>. <u>occidentalis</u> Wats. (Mountain Ash)

Smith's transcription of Gitksan name: skanklingit, klingit meaning the fruit. John Fowler said there were only a few trees on the terraces of the valleys but many on the mountains.

Modern spelling of Gitksan name: s<u>g</u>anhlingit (plant), hlingit (fruit, literally, 'slave')[152]

Among the Gitksan, according to John Fowler, September 16, 1925, the fresh raw fruit was crushed and eaten as a strong physic, never boiled or cooked in any way. The tree had no further use, the roots, and bark being useless and the tree having no myth. Luke Fowler, June 8, 1926, said that the wood was used for axe handles (see Figure 45).

---

[152] Hlingit means "slave," and it is borrowed from the Tlingit word of similar shape, which means 'man, person' in that language. The Tsimshianic-speaking peoples used to take slaves regularly from among their Tlingit neighbours. Still today, a grandfather might sometimes call his grandchild hlgu hlingit, 'little slave,' an affectionate term.

Figure 45. Sorbus sitchensis (Sitka mountain-ash)

(89) <u>Spiraea douglasii</u> Hook. ssp. <u>menziesii</u> (Hook.) Calder & Taylor (Hardhack, Pink Spiraea or Menzies' Spirea) and possibly <u>S</u>. <u>douglasii</u> Hook. ssp. <u>douglasii</u> (Hardhack, or Douglas' Spirea)

Identified by Smith as: (94) (118) (137) <u>Spiraea Douglasii</u> Hook. var. <u>Menziesii</u> Presl. (Hardhack) (a bush with a spike the colour of raspberries and cream) ([94] from Kitwanga, B.C., June 21, 1926)

Smith's transcription of Gitksan name: hisganthoot or his gant hoot, his means similar (Bob Robinson)

Modern spelling of Gitksan name: hisgantxwit ("it's a pretend/false-stick")[153]

Among the Gitksan, according to Abraham Fowler, June 21, 1926, and Bob Robinson, October 3, 1926, this plant was of no use.

Salicaceae (Willow Family)

(90) <u>Populus balsamifera</u> L. ssp. <u>trichocarpa</u> (T. & G.) Brayshaw (Black Cottonwood) and possibly <u>P</u>. <u>balsamifera</u> L. ssp. <u>balsamifera</u> (Balsam Poplar)

Identified by Smith as: <u>Populus trichocarpa</u> T. & B. (Black Cottonwood)

Smith's transcription of Gitksan name: am mel, "good for canoe," mael, "canoe"

Modern spelling of Gitksan name: am'mel (WG), am'mal (EG) (literally, 'good for-canoe')

Among the Gitksan, according to Luke Fowler, May 24, 1926, the inner bark of the black cottonwood was scraped from the trunk of young trees for food. It was eaten at the end of May and in June the same as with jack pine and aspen.

The wood was used for dugout canoes, smoke houses and firewood. It was not used as a fire drill spindle or hearth as among the Nuxalk, according to Luke Fowler, June 15, 1926; the hearth was made of cedar or willow. The new leaves were used in a bath as among the Nuxalk. Fresh cottonwood branches without the leaves were seen in use at Hazelton August 13, 1926 as a carpet on which salmon were being cut up apparently to protect them from the underlying sand.

---

[153]This is probably a nonce form and is the same term given for <u>Artemisia michauxiana</u>.

The gum of the buds was used in a hair perfume by young ladies. The young buds were boiled and put in bear grease and kept in a bone with a hole in the side. The roots and seeds were of no use (see Figure 46).

Figure 46. Populus balsamifera (black cottonwood)

(91) *Populus tremuloides* Michx. (Trembling Aspen)

Identified by Smith as: *Populus tremuloides* Michx. (Trembling Aspen)

Smith's transcription of Gitksan name: am gäst, am gost, am gawst or am gäst (gawat), am meaning good and gäst maple; that is good for maple, very fine maple, or better than maple

Modern spelling of Gitksan name: amk'ooxst (literally, 'good for-maple')

Among the Gitksan, according to Luke Fowler, May 24, 1926, the inner bark of the aspen like that of the cottonwood and scrub pine was scraped from the peeled trunk of the tree and eaten. The wood was used for firewood and for masks, it being white. The bark on the roots was chewed or mashed and put in cuts. The bark of the trunk was boiled by itself and the decoction drunk as a purgative, not an emetic. The trunk was too small to make into a canoe and the wood was not used for paddles.

(92) *Salix sitchensis* Sanson ex Bong. (Sitka Willow)

Identified by Smith as: (48) *Salix* sp. (probably *S. sitchensis* Bong.) (Willow)[154] (from gravel terrace near New Hazelton, B.C.)

Smith's transcription of Gitksan name: washantsmilt, beaver willow tree, washan meaning willows and tsmilt meaning beaver[155]

Modern spelling of Gitksan name: 'waasanhl ts'imilix (literally, 'willow of beaver,' or 'beaver-willow')

According to Luke Fowler, May 24, 1926, it is small, comes from near mountains and is the best beaver food. The beaver camps near it and never leaves the area where it grows.

(93) *Salix* sp. (Willow)

Identified by Smith as: (95) *Salix* sp. (Willow)[156]

Smith's transcription of Gitksan name: washan

Modern spelling of Gitksan name: 'waasan

---

[154] Elsewhere, Smith (1929:53, 54) identified "*Salix lasiandra* Benth. (Willow)" (syn.: *S. lucida* Muhl. [shining willow, Pacific willow and tail-leaved willow]) and "*S. scouleriana* Hook. (Willow)" (i.e., *S. scouleriana* Barratt ex Hook. [Scouler's willow]) and described their medicinal uses among the Nuxalk.

[155] The beaver (*Castor canadensis* Kuhl) is known in Gitksan as ts'imilix.

[156] Smith also associated the number 95 with *Galium boreale* (see Appendix 4, #44) but the information presented here applies to willow.

There is only one kind of willow in the Gitksan country besides the beaver willow, according to Luke Fowler. There are over thirty kinds in Southern British Columbia and probably many in the Gitksan country.

Among the Gitksan, according to Luke Fowler, May 24, 1926, willow was good for firewood; the roots for fire drill spindles, but the bark was of no use. Fire drill spindles were not made of cottonwood. The hearth was made of cedar or willow.

Brown, a Gitksan Indian at Hazelton, B.C. said willow bark was used for string for drying salmon (June 19, 1927).

Editorial comments: Elsewhere Smith stated that thimbleberry leaves were tied with willow to make a ball to play with.

Santalaceae (Sandalwood Family)

(94) Geocaulon lividum (Richards.) Fern. (Bastard Toad-flax, or Northern Comandra)
Identified by Smith as: (5) Comandra livida Rich. (Bastard Toad-flax) (from first terrace,
    south of Skeena River, near Kitwanga, B.C., August 23, 1925)
Smith's transcription of Gitksan name: măĕtswĭgĕt (măetsweeget)
Modern spelling of Gitksan name: maa'yts 'Wii Get (WG), maa'yts 'Wii Gat (EG) ("it's 'Wii
    Get/'Wii Gat's berry")

Among the Gitksan, according to John Fowler, August 27, 1925, this plant was of no use, but there is a long story of Wiget[157], which Mr. C. Marius Barbeau collected and this story relates how Wiget was the first man. Wiget liked to eat salmon and the salmon came out of the water on to the ground for him to eat them. He boiled and ate the salmon and he ate all the berries even this kind which are not eaten by the Gitksan. After eating all the berries he would move on to another place to eat.

Scrophulariaceae (Figwort Family)

(95) Castilleja miniata Dougl. ex Hook. (Scarlet or Common Red Paintbrush)

---

[157] Smith noted that "according to the Indians as Adam was according to white people also that this first man (Wiget)." This term is properly spelled 'Wii Get (WG) or 'Wii Gat (EG), although Smith's spelling has been retained here except in the case of retranscribed plant names.

Identified by Smith as: (93) Castilleja miniata Dougl. (Paint-Brush) (from Kitwanga, B.C., June 21, 1926)

Smith's transcription of Gitksan name: ik lē ām chŭk (ik clay am chock) good for nose bleed. Iklē means blood; am, good; and chŭk, nose. As there are two plants called by this name, this one is called short, the other 136 (40) [i.e., Aquilegia formosa ssp. formosa] tall.

Modern spelling of Gitksan name: ihlee'em ts'ak (literally, 'bleeding nose')

Among the Gitksan, according to Luke Fowler, May 25, 1926, this plant was boiled, roots and all, and the decoction drunk for nose bleed, bleeding, stiff lungs, bad eyes, and lame back, possibly caused by kidney trouble. It is a purgative and diuretic.

Among the Gitksan, according to Abraham Fowler, June 21, 1926, the seeds of this plant were boiled and the decoction drunk for coughs (see Figure 47).

Figure 47. Castilleja miniata (common red paintbrush)

Solanaceae (Potato or Nightshade Family)

(96) †Nicotiana tabacum L. (Tobacco)
Identified by Smith as: Nicotiana tabacum L. (Tobacco)
Gitksan name: (no name recorded by Smith)[158]

There was no tobacco in the old days according to John Fowler, August 20, 1925.[159]

Urticaceae (Nettle Family)

CAUTION—THE FOLLOWING SPECIES IS CAPABLE OF CAUSING DERMATITIS!

(97) Urtica dioica L. ssp. gracilis (Ait.) Seland. (Stinging Nettle)
Identified by Smith as: (29) Urtica Lyallii Wats. (Western Nettle) (Seen at Kitwanga, B.C., August 9, 1925.)
Smith's transcription of Gitksan name: stetsh
Modern spelling of Gitksan name: sdetxs (WG), sdatxs (EG)

Among the Gitksan, according to Abraham Fowler, August 9, 1925, nettle fibre was not used for making rope. It is well known that it is used for making net twine among the Nuxalk.

Among the Gitksan, according to Luke Fowler, May 24, 1926, nettles were gathered in August and the fibre was made into thread for nets and for the short cross element in pack straps. If gathered later the plant would be too hard. Some people use it for medicine. The stalks and roots were boiled and the decoction drunk for lungs or bladder, when a man spit blood, and for any sickness. Nettles were not used (as a counterirritant) to sting a person according to Luke Fowler, May 24, 1926 although they are so used among the Nuxalk (see Figure 48).[160]

---

[158] Hindle and Rigsby (1973:48) recorded the term mi'yan for tobacco.

[159] Nicotiana tabacum is a post-contact introduced species used as a fumitory and masticatory.

[160] Elsewhere, Smith (1927) reported that "The Gitksan Indians spun the fibre of this plant into cord used chiefly for fish nets, but also for such purposes as to tie boxes and blankets and for one element in woven tump lines. The entire plant was boiled and the decoction drunk for many ailments, including hemorrhage from the mouth and lungs, and bladder troubles." Guédon (1973) has also reported that stinging nettle roots were boiled and used as a poultice, that stinging nettle was never taken internally, and that this plant was always used in conjunction with another medicine. She further reported that leaves of cow-parsnip were used to counteract the pain of being stung by stinging nettles.

Figure 48. Urtica doica (stinging nettle)

Valerianaceae (Valerian Family)

(98) Valeriana dioica L. (Marsh Valerian)
Identified by Smith as: (69) Valeriana septentrionalis Rydb. (Valerian) (not in Henry's flora)
(from Kitwanga, B.C.)
Smith's transcription of Gitksan name: ishshimskōks (ish shim skooks)
Modern spelling of Gitksan name: (unrecognizable)[161]

Among the Gitksan, according to Luke Fowler, June 15, 1926, this plant was used for the best perfume. The whole plant was put in bear grease and the grease was used on the face and hair.

Angiosperms (Flowering Plants), Monocotyledons

Araceae (Arum Family)

(99) ?Calla palustris L. (Wild Calla)
Identified by Smith as: (51) Calla palustris L. (Water Arum, Calla Lily)
Smith's transcription of Gitksan name: shiĕn (she en)[162]
Modern spelling of Gitksan name: (unrecognizable)

The root tastes like a banana. From pond near Hazelton, B.C., May 25, 1926. Among the Gitksan, according to Luke Fowler, May 25, 1926, a decoction made of the roots of this plant by boiling them well was drunk for cleaning the eyes of the blind, for hemorrhage of the lungs and from the mouth, for short breath and flu which was unknown until recently (see Figure 49).

---

[161] It is not possible to reconstitute the whole of this form from Smith's spelling, but it appears to be an incorporated compound verb that has is 'smell' in its initial element (is-im). Its meaning would be "smell like skooks," whatever that might be.

[162] While C. palustris does occur within Gitksan territory, this term may actually refer to Potentilla anserina L. ssp. pacifica (Howell) Rousi (silverweed), a plant known among other Tsimshianic groups who used the roots as food by the cognates siyin (Coast Tsimshian) and syaan (Southern Tsimshian). The only comparable Gitksan word is syan, "to be tainted from having transgressed." Gottesfeld and Anderson (1988) have also reported hisgahldaatsxw for wild calla, which they translate as "resembles yellow pond lily." This term likely should be hisgahldaatsxwit, "it's a pretend-yellow pond lily."

Figure 49. Calla palustris (wild calla)

Liliaceae (Lily Family)

(100) <u>Allium</u> <u>cernuum</u> Roth in Roem. (Nodding Onion)
Identified by Smith as: <u>Allium</u> sp. (Wild Onion)
Smith's transcription of Gitksan name: jenshagaʳk
Modern spelling of Gitksan name: ts'enksa g̱aak̲ (WG), ts'anksa g̱aak̲ (EG) (literally, 'armpit of raven,' or 'raven-armpit')[163]

Among the Gitksan, according to Luke Fowler, May 25, 1926, the wild onion was eaten raw, and the entire plant was boiled in with rabbit[164] or with meat or fish of any kind.

(101) <u>Clintonia</u> <u>uniflora</u> (Schult.) Kunth (Queen's Cup, or Blue-bead Clintonia)
Identified by Smith as: (100) <u>Clintonia</u> <u>uniflora</u> Kunth. (Queen's Cup) (from Kitwanga, B.C., June 21, 1926)
Smith's transcription of Gitksan name: häwbush Wīget (häbush means spoon. Wīget is the name of the first man and is a name still in use. wī means big, and get, man.)
Modern spelling of Gitksan name: hoobixs 'Wii Get (WG), hoobixs 'Wii Gat (EG) (literally, "Wii Get's ['Wii Gat's] spoon')[165]

Among the Gitksan, according to Luke Fowler, June 8, 1926, this plant was of no use, although a leaf is said to have been used as a spoon by Wiget, the first man. The story runs as follows: "Wiget came to an invisible town made of air, the location of which is unknown. He went to one of the houses but there were no people in it, although he could hear people. He heard them laugh. They spoke to him, but he could not see them and so he called them air people. He saw a salmon, a kettle-basket, hot stones and a leaf of Queen's Cup which was a spoon. The salmon of its own accord went into the kettle-basket. The hot stones followed. The salmon when it was cooked went into a trough-shaped wooden dish and the Queen's Cup leaf spoon followed. Wiget ate the salmon with this leaf spoon."

Among the Gitksan, according to Abraham Fowler, June 21, 1926, this plant was of no use.

---

[163] The raven in question here is common raven (<u>Corvus</u> <u>corax</u> L.) and gaak̲ is the common noun for this species, not the proper noun name of the Myth Age character, Raven.
[164] Here, "rabbit" refers to <u>Lepus</u> <u>americanus</u> (Erxleben), snowshoe hare, known in Gitksan as ga̱x.
[165] 'Wii Get, or 'Wii Gat ("Giant," literally, 'Big Man') is the main transformer-trickster character in Gitksan oral literature.

(102) <u>Disporum</u> <u>hookeri</u> (Torr.) Nichols. var. <u>oreganum</u> (S. Wats.) Q. Jones (Hooker's Fairybells)
Identified by Smith as: (85) <u>Disporum</u> <u>oreganum</u> B. & Hook. (Fairy-Bells)
Smith's transcription of Gitksan name: skān căp căp (skan cap coip) (skan meaning plant)[166]
Modern spelling of Gitksan name: sgangapk'oyp

Among the Gitksan, according to Bob Robinson, July 24, 1926, this plant was of no use and had no story about it.

(103) ?<u>Disporum</u> <u>trachycarpum</u> (S. Wats.) Benth. & Hook. (Rough-fruited Fairybells) or perhaps some species of <u>Streptopus</u>
Identified by Smith as: (88) ?<u>Smilacina</u> sp. ("a sort of branching Solomon seal")
Smith's transcription of Gitksan name: skan sāxstoo′lŭ, skan meaning plant; the fruit was called sāxstoo′lŭ
Modern spelling of Gitksan name: sgansaxsduu'lxw (literally, 'plant [undefined]'), saxsduu'lxw (fruit [undefined])[167]

Among the Gitksan, according to Bob Robinson, July 24, 1926, this plant was of no use—not even the roots—and it had no story about it.

(104) <u>Fritillaria</u> <u>camschatcensis</u> (L.) Ker-Gawl. (Northern Rice-root, or Riceroot Fritillary)
Identified by Smith as: (87) (88) <u>Fritillaria</u> <u>kamtschatcense</u> Ker. (Fritillary, Rice Root)[168]
Smith's transcription of Gitksan name: koss or kāsx (not kosh)
Modern spelling of Gitksan name: gasx (literally, 'be bitter' [singular adjective form])

Among the Gitksan, according to Robert A. Sampare, May 26, 1926, the rice-like bulbs were boiled for food and served with eulachon oil, salmon oil, bear grease, or groundhog grease.

According to Abraham Fowler, May 29, 1926, the rice-like bulbs are ready for gathering June 15th. The women washed them clean, spread them on a mat and dried them for

---

[166]The term that Smith reported for <u>D</u>. <u>hookeri</u> var. <u>oreganum</u> is equivalent to the term he recorded for <u>Cornus</u> <u>canadensis</u> (gapk'oyp) with the addition of the word glossed as "plant." It is possible that <u>D</u>. <u>hookeri</u> var. <u>oreganum</u> is named in recognition of some perceived similarity or relationship to <u>C</u>. <u>canadensis</u> or, that this plant may have been mistaken for <u>C</u>. <u>canadensis</u>. <u>Cornus</u> <u>canadensis</u> is also known locally as "snowberry" or "popberry."
[167]These represent postulated forms with undefined meanings.
[168]Although Smith gave two collection numbers for this species the actual collection number is 88.

winter use. One woman would easily and quickly gather and dry as much as five or six hundred pounds (see Figure 50).

Editorial comments: Smith also wrote that riceroot bulbs were sometimes served with western hemlock cambium and one or more types of animal oil. The bulbs may also have been occasionally dried. This information seems to have come from Luke Fowler (on June 15, 1926).

Figure 50. Fritillaria camschatcensis (northern rice-root)

(105) _Smilacina_ _racemosa_ (L.) Desf. var. _amplexicaulis_ (Nutt. ex Baker) S. Wats. (False Solomon's-seal)

Identified by Smith as: (13) _Smilacina_ _racemosa_ L. (False Solomon's Seal) (Kitwanga, B.C., August 23, 1925)

Smith's transcription of Gitksan name: scon.c⁹ōts (sconcoats), skan gaots or sconcoots, gōts, gaots or coots meaning the fruits, skan meaning plant

Modern spelling of Gitksan name: (unrecognizable) (literally, 'plant [undefined]')[169]

Among the Gitksan, according to John Fowler, August 27, 1925, the fruits were eaten raw, and boiled, but were not dried or put on racks in the form of cakes. The fruit was boiled and kept in a small box without oil by some old women. According to Luke Fowler, May 24, 1926, the fruits were put in with black berries and huckleberries[170] and dried in cakes in the sun or a fire or eaten raw with grease. Bears eat the berries, too. According to Bob Robinson, July 24, 1926, the fruit was eaten raw, sometimes after being boiled, but was not dried in cakes. The Gitksan women picked the fruit in the autumn, boiled it, and kept it in a box for winter use. It was served after eating salmon and was best with eulachon oil. Poor people and those in a hurry ate it without oil. The skin of the fruit was swallowed but the seeds were spit out.

According to John Fowler, August 27, 1925, the roots were boiled and the resulting decoction was used as a very strong medicine for rheumatism. The bark and leaves were of no use. According to Luke Fowler, May 24, 1926, the roots were boiled for medicine with a clearing or purgative effect used for sore back, kidney ailments, and rheumatism. According to Bob Robinson, July 24, 1926, the roots were used for a cut medicine. They were cut with an axe and mashed up like dough then put in cuts and bound on. This was a fine medicine.

---

[169]This proper transcription of this term may be sgank'ots, where k'ots may refer to the fruits. Note that this plant has also been referred to curiously as "Indian glads" (perhaps through an erroneous association between this plant and species of _Gladiolus_) and otherwise incorrectly identified as "_Smilacina_ _amplexicaulus_" (People of 'Ksan 1980:70, 71).

[170]Here Smith refers to "black berries" and "huckleberries." Elsewhere he refers to "black huckleberries (mae or my)," "black bilberries," and "high bush blueberries." All of these terms are suggestive of species of _Vaccinium_. However, within the Tsimshianic languages there is much confusion regarding the indigenous terms and botanical identities of _Vaccinium_ spp.—a fact especially true for Gitksan. The "black berries" referred to by Smith apparently are not blackberries (also known as blackcaps or black raspberries, i.e., _Rubus_ _leucodermis_ Doug. ex T. & G.). This may be surmised because in association with red raspberries Smith wrote that "Luke Fowler said the Gitksan had no black raspberries, but that he had heard of them." Neither does "black berries" refer to trailing raspberry (_Rubus_ _pubescens_), since these were said to be uncommon and were not cooked, unlike the "black berries" which Smith reported were dried in cakes. Smith also referred to "black huckleberries (mae or my)" and "black bilberries." These common names may refer to _Vaccinium_ _membranaceum_ (black huckleberry), a species known in other Tsimshianic languages as tú?utsgm m̓á'ı or tú?u'tsgm m̓á'ı, literally, 'black berry' (Southern Tsimshian) and miigaan (Nisga'a). The Gitksan term associated with "black huckleberries"—mae or my—is simply the term for '(any) berry (or fruit).'

Smith also referred to a species identified as "high bush blueberry," "huckleberries" and skan mı gan (sganmiigan) or migaʳan (miigan), possibly _Vaccinium_ _alaskaense_ (Alaskan Blueberry) and/or _V._ ovalifolium (oval-leaved blueberry).

(106) <u>Smilacina</u> <u>stellata</u> (L.) Desf. (Star-flowered False Solomon's-seal)

Identified by Smith as: <u>Smilacina</u> <u>stellata</u> (L.) Desf. (Star-flowered False Solomon's-seal)

Smith's transcription of Gitksan name: hish gān gō sh^u, hĭshgāngōch^u, hish meaning "funny" or peculiar (gān gō ch, "funny")

Modern spelling of Gitksan name: (unrecognizable)[171]

Among the Gitksan, according to Luke Fowler, June 8, 1926, this plant was of no use and had no name.

(107) <u>Streptopus</u> <u>amplexifolius</u> (L.) DC. in Lam. & DC. var. <u>americanus</u> Schult. and/or <u>S</u>. <u>amplexifolius</u> (L.) DC. in Lam. & DC. var. <u>chalazatus</u> Fassett (Clasping Twistedstalk)

Identified by Smith as: (44) <u>Streptopus</u> <u>amplexifolius</u> DC. (Twisted-stalk) (from near Hazelton, B.C., 1926)

Smith's transcription of Gitksan name: stōlĭks (stolicks)

Modern spelling of Gitksan name: xsduu'lixs[172]

Among the Gitksan, according to Luke Fowler, May 24, 1926, this plant was of no use, although the fruits are like raisins or grapes.

CAUTION—THE FOLLOWING SPECIES IS HIGHLY TOXIC!

(108) <u>Veratrum</u> <u>viride</u> Ait. ssp. <u>eschscholtzii</u> (A. Gray) Löve & Löve (Indian or Green False Hellebore);

Identified by Smith as: <u>Lysichiton</u> <u>kamtschatcense</u> Schott. (Skunk Cabbage)[173]

Smith's transcription of Gitksan name: The leaves were called skan chĭcks and the roots melwash^u;[174] Skan means plant

---

[171] The term reported by Smith may be hissgank'ots ("it's a pretend/false false Solomon's-seal").

[172] This form is not known from other studies as a Gitksan plant name but it represents the best reconstitution of Smith's term which may be derived from xsduu'l, 'tears.'

[173] The following information does not pertain to <u>Lysichiton</u> <u>americanum</u> Hult. & St. John (skunk-cabbage; Araceae), known to the Gitksan as hinak. Smith erroneously specified skunk-cabbage in connection with these data here and elsewhere (Smith 1929:52, 53). The information presented here actually refers to <u>Veratrum</u> <u>viride</u> ssp. <u>eschscholtzii</u>, , a highly toxic plant known to have been used with caution by various First Nations of British Columbia (cf. Gottesfeld and Anderson 1988). The following text has been corrected to specify Indian hellebore where Smith originally erroneously specified skunk-cabbage. Likewise, all other erroneous references to skunk-cabbage in this document have been edited to specify Indian hellebore.

[174] Smith's "melwash^u" represents an inaccurate rendering of melgwasxw (WG), malgwasxw (EG), the antipassive nominalization of the singular transitive active verb malkw ("burn"), meaning "something that has burned."

Modern spelling of Gitksan name: sgants'iks (leaves or plant); melgwasxw (WG), malgwasxw (EG) (root)

Among the Gitksan, according to Luke Fowler, May 24, 1925, the root of the Indian hellebore was well mashed, wet, and applied for blood poison and boils. It killed the poison, brought the boils to a head, and was used with other plant products as a medicine for boils and ulcers, and for hemorrhage of the lungs. The root mixed with many other plant products was also used for rheumatism. Smoke of the roots was inhaled for bad dreams, flu and rheumatism, but would kill a medicine man, because it kills his medicine. The roots were not used for food as they were poison; the water from them if drunk would stop the tongue and all talk as it is poison. The leaves were used to sit or lie on when taking a sweat bath for rheumatism and were put over the lame places.

For a medicine the roots were mashed together with the large round green rootstock, but not the rootlets of a fern called demk (demtx [WG], damtx [EG]), bark of balsam fir and devil's club, and a little gum of scrub pine or spruce and the mass was warmed a little and applied to boils and ulcers. It caused them to come to a head. If this medicine was not applied they suppurated. It was also put on the chest for hemorrhage of the lungs, but sometimes the patient was too far gone (see Figure 51).

Editorial comments: Smith also wrote that the leaves of the plant discussed above were not used (as among the Nuxalk) for making shelters. or temporary drinking cups. It should be noted that the leaves in question as used among the Nuxalk are those of skunk-cabbage, rather than those of Indian hellebore. From data that Smith associated with Platanthera obtusata (#109), it would seem that the leaves of skunk-cabbage were used by the Gitksan as they were among other Northwest Coast groups—to line berry-drying racks.

Figure 51. Veratrum viride (Indian hellebore)

Orchidaceae (Orchid Family)

(109) <u>Platanthera</u> <u>obtusata</u> (Banks ex Pursh) Lindl. (One-leaved Rein Orchid)
Identified by Smith as: (131) <u>Habenaria</u> <u>obtusata</u> Richards (Rein Orchid)
Smith's transcription of Gitksan name: his hĭnnāk hŭt, his meaning "resembles," hĭnnāk means
a plant the leaves of which may be used on which to dry berries.[175]
Modern spelling of Gitksan name: hishina<u>k</u>xwit ("it's a pretend/false-skunk-cabbage")[176]

Among the Gitksan, according to Bob Robinson, October 3, 1926, this plant was of no use. It is too small, and the roots are of no use.

Poaceae (syn.: Gramineae, Grass Family)

(110) ?<u>Elymus</u> <u>glaucus</u> Buckl. (Blue Wildrye) or <u>E</u>. <u>trachycaulus</u> (Link) Gould in Shinners
(Slender Wheatgrass)
Identified by Smith as: (65) Bunch Grass (?<u>Agropyron</u> sp.)
Smith's transcription of Gitksan name: hăbăshh^u or habash, a white sharp long grass that
grows in meadows
Modern spelling of Gitksan name: habasxw (literally, 'something that covers')[177]

Among the Gitksan, according to Luke Fowler, June 8, 1926, bunch grass was used for socks or stuffing around the feet in moccasins, for bedding for babies, and as a covering for the ground where the people sat around the fire, there formerly being no wooden floors.

(111) Unidentified "Grass" (Poaceae?)
Identified by Smith as: "grass"
Gitksan name: (no name recorded by Smith, but habasxw is the general word for 'grass')

<u>Editorial comments</u>: Smith wrote that lupine roots were cooked in a hole in the earth covered with one or more unidentified types of grass.

---

[175] This term incorporates the name for <u>Lysichiton</u> <u>americanum</u> and probably represents a nonce form. Also note that a different term said to have a similar meaning was reported by Smith for <u>Geranium</u> <u>richardsonii</u>.
[176] This is probably a nonce form but it serves to validate the existence of the term hina<u>k</u> which clearly refers to <u>Lysichiton</u> <u>americanum</u>, the leaves of which were traditionally used to line berry drying racks among many Northwest Coast cultures.
[177] This is the general word for 'grass,' and it is an antipassive nominalization of the singular transitive verb hap, 'cover.'

Unidentified Species

(112) Unidentified Plant
Smith's transcription of Gitksan name: jēuk
Modern spelling of Gitksan name: (unrecognizable)[178]

<u>Editorial comments</u>: Of the plant identified as jēuk, Smith wrote only that it was regarded as similar (in having a pungent smell) to <u>Ribes hudsonianum</u> (#61) and <u>Rubus pubescens</u> or <u>R. pedatus</u> (#86). However, in contrast to those other two species, jēuk was said to grow on mountains.[179]

Discussion, Summary, and Conclusions

Smith's report on Gitksan ethnobotany contains references to as many as 165 plants and fungi, including up to 112 species that reportedly have Gitksan names and cultural roles (see Tables 1-8). The named and culturally significant species are representative of as many as five families of fungi (including lichens), four families of pteridophytes, three families of gymnosperms and 33 families of angiosperms (including 29 families of dicotyledons and four families of monocotyledons)

---

[178]This proper transcription of this term may be jeekw.
[179]The term jēuk seems similar to another word recorded by Smith apparently in reference to the nut of <u>Corylus cornuta</u> var. <u>cornuta</u> (beaked hazelnut), i.e., jekhh.

Table 1. Summary of botanical species used as food or in food-related applications among the Gitksan.[180]

Entry Number	Species[181]	Food or Food-related Application(s)
9	?"Dryopteris austriaca complex" (wood ferns)	(rootstock) eaten
17	Abies amabilis (amabilis or Pacific silver fir) and A. lasiocarpa (subalpine or alpine fir)	cambium eaten
18	Picea x lutzii (Hybrid Sitka spruce)	cambium eaten
19	Pinus contorta (lodgepole pine)	cambium eaten
23	Angelica genuflexa (kneeling angelica)	plant eaten to prevent bears from smelling hunters; hollow stems used as drinking tubes
24	Heracleum lanatum (cow-parsnip)	peeled stems eaten; hollow stems used as drinking tubes
35	Alnus rubra (red alder) and possibly also A. tenuifolia (mountain alder)	branches used as undermat upon which salmon were cleaned
38	Corylus cornuta (beaked hazelnut)	leafy branches used as undermat upon which salmon were cleaned
42	Sambucus racemosa (coastal red elder)	fruits eaten
44	Viburnum edule (highbush cranberry)	fruits eaten
47	Cornus stolonifera (red-osier dogwood)	withes used as undermat upon which salmon were cleaned
49	Shepherdia canadensis (soopolallie)	fruits eaten
50	Arctostaphylos uva-ursi (kinnikinnick)	fruits eaten; leaves mixed with tobacco and smoked
51	Ledum groenlandicum (Labrador tea)	decoction of leaves drank as beverage
53	Oxycoccus oxycoccus (bog cranberry)	fruits eaten
54	Vaccinium ?alaskaense (Alaskan blueberry) and/or V. ?ovalifolium (oval-leaved blueberry)	fruits eaten
55	Vaccinium caespitosum (dwarf blueberry)	fruits eaten
56	Vaccinium ?membranaceum (black huckleberry)	fruits eaten
57	Lathyrus ochroleucus (creamy peavine)	used to wipe salmon or meat
58	Lupinus arcticus (arctic lupine)	roots and fruits eaten
60	Ribes hudsonianum (northern blackcurrant)	fruits eaten
62	Ribes oxyacanthoides (northern gooseberry)	fruits eaten
66	Epilobium angustifolium fireweed)	inner part of young shoots eaten
67	Rumex acetosella (sheep sorrel)	stems and leaves eaten
70	Aquilegia formosa (red columbine)	nectar from flowers eaten
75	Amelanchier alnifolia (saskatoon)	fruits eaten
77	Crataegus douglasii black hawthorn)	fruits eaten
79	Fragaria virginiana (wild strawberry)	fruits eaten
80	Malus fusca (Pacific crab apple)	fruits eaten
81	?Prunus emarginata (bitter cherry) or, more likely, P. pensylvanica L. (pin cherry)	fruits eaten
82	Prunus virginiana (choke cherry)	fruits eaten

[180] The cultural role data that are summarized here and in the subsequent tables are based on a discussion presented by Turner (1988) on the cultural significance of plants among some First Nations groups of British Columbia.
[181] For the sake of brevity, neither infraspecific taxa nor alternative English plant names have been included here or in the subsequent tables.

83	Rosa acicularis (prickly rose) and/or R. nutkana (Nootka rose)	hips eaten
84	Rubus idaeus (red raspberry)	fruits eaten
85	Rubus parviflorus (thimbleberry)	fruits eaten; leaves used as undermat upon which berries were dried
86	Rubus pubescens (trailing raspberry) or possibly R. pedatus (five-leaved creeping raspberry)	fruits eaten
87	Rubus spectabilis (salmonberry)	fruits eaten
90	Populus balsamifera (black cottonwood, balsam poplar)	cambium eaten; branches used as undermat upon which salmon were cleaned; wood used as fuel in smokehouse
91	Populus tremuloides (trembling aspen)	cambium eaten
100	Allium cernuum (nodding onion)	entire plant eaten
104	Fritillaria camschatcensis (northern rice-root)	bulbs eaten
105	Smilacina racemosa (false Solomon's-seal)	fruits eaten (seeds spit out)
111	unidentified "grass"	used to cover cooking pits

Table 2. Summary of botanical species regarded as animal food among the Gitksan.		
Entry Number	Species	Associated Zoological Species
7	unidentified "mosses"	Ursus spp. (bears)
23	Angelica genuflexa (kneeling angelica)	frogs and/or toads
27	Aralia nudicaulis (wild sarsaparilla)	Ursus americanus (black bear)
43	Symphoricarpos ?albus (common snowberry)	(unidentified) "sparrow"?
46	Cornus canadensis (bunchberry)	"chipmunk" (?Tamias amoenus, (yellow-pine chipmunk)
47	Cornus stolonifera (red-osier dogwood)	Ursus americanus (black bear) and U. arctos (grizzly bear)
68	Actaea rubra (baneberry)	Ursus americanus (black bear)
72	Ranunculus abortivus (kidney-leaved buttercup)	Lontra canadensis (river otter)

Table 3. Summary of botanical species used as material for technological and other applications among the Gitksan.

Entry Number	Species	Technological Application(s)
6	Sphagnum spp. (peat moss)	used for diapers
10	Equisetum arvense (common or field horsetail)	abrasive stalks used to polish wood
14	Chamaecyparis nootkatensis (yellow cedar)	wood used for paddles
15	?Juniperus communis (common or ground juniper)	wood used for bows and as firewood
16	Thuja plicata (western redcedar)	wood used as lumber, and for canoes, totem poles and other applications; inner bark used in basketry and other applications; limbs used as rope; bark used for house roofs
17	Abies amabilis (amabilis or Pacific silver fir) and A. lasiocarpa (subalpine or alpine fir)	wood used for lumber, firewood, and for making snowshoes; bark used for roofs
18	Picea x lutzii (Hybrid Sitka spruce)	wood used for handles of salmon gaffs, for wings of salmon traps, and as firewood; bark used for roofs; roots used in basketry and construction of salmon traps
19	Pinus contorta (lodgepole pine)	wood used in salmon traps and for canoe poles, and for torches and firewood
20	Tsuga heterophylla (western hemlock)	wood used for wedges, salmon traps, spear points, canoe poles, and firewood
21	Taxus brevifolia (western or Pacific yew)	wood used for adze handles
22	Acer glabrum (Douglas or Rocky Mountain maple)	inner bark used in basketry and to make pack sacks and woven mats; wood used for spoons, snowshoes, rattles, and paddles
26	Apocynum androsaemifolium (spreading dogbane)	fibre used for cordage used in nets, pack-straps, and animal snares
35	Alnus rubra (red alder) and possibly also A. tenuifolia (mountain alder)	decoction of bark used as red dye; wood used for paddles, adze handles, masks, and firewood
36	Alnus tenuifolia (mountain alder) or possibly A. crispa (green or Sitka alder)	wood used for spoons and as firewood
37	Betula papyrifera (paper birch)	wood used for firewood; bark used for pails, basins, torches, and spoons
38	Corylus cornuta (beaked hazelnut)	leafy branches used as undermat upon which salmon were cleaned
43	Symphoricarpos ?albus (common snowberry)	hollow stems used as pipe stems
44	Viburnum edule (highbush cranberry)	branches twisted into rope
47	Cornus stolonifera (red-osier dogwood)	withes used for rope
66	Epilobium angustifolium fireweed)	fibre used for cordage used for fish nets, pack straps, and tying blankets and boxes
75	Amelanchier alnifolia (saskatoon)	wood used for adze handles and arrows
77	Crataegus douglasii black hawthorn)	wood used for adze and axe handles, and fishing hooks
80	Malus fusca (Pacific crab apple)	wood used for adze and axe handles
82	Prunus virginiana (choke cherry)	wood used for adze handles
83	Rosa acicularis (prickly rose) and/or R. nutkana (Nootka rose)	wood used for arrow points
88	Sorbus sitchensis (Sitka mountain-ash) and/or S. scopulina (western mountain-ash)	wood used for axe handles

90	Populus balsamifera (black cottonwood, balsam poplar)	wood used for dugout canoes, smoke houses, and firewood; buds and leaves used as perfume agent (leaves in bath, buds on hair)
91	Populus tremuloides (trembling aspen)	wood used for masks, and as firewood
93	Salix sp. (willow)	wood used as firewood; roots used for fire drill spindles; bark used to hang drying salmon
97	Urtica dioica (stinging nettle)	fibre used for cordage used in nets and pack straps
98	Valeriana dioica (marsh valerian)	plant used as perfume agent mixed with bear grease and used on face and hair
110	?Elymus glaucus (blue wildrye) or ??E. trachycaulus (slender wheatgrass)	used for socks or stuffing inside moccasins, for baby bedding, and to cover ground where people sat around the fire

Table 4. Summary of botanical species used as medicinal applications among the Gitksan.

Entry Number	Species	Medicinal Application(s)
2	Inonotus obliquus (cinder conk)	smouldering fungus used to treat rheumatism
8	?Athyrium filix-femina (lady fern), ?Pteridium aquilinum (bracken fern)	rootstock combined with other medicines to treat boils, ulcers, and hemorrhage of the lungs
15	?Juniperus communis (common or ground juniper)	decoction of entire plant used as purgative and diuretic for various complaints including hemorrhage at the mouth and kidney trouble
17	Abies amabilis (amabilis or Pacific silver fir) and A. lasiocarpa (subalpine or alpine fir)	cambium eaten for constipation; pitch of young cones used as purgative and diuretic for tuberculosis and gonorrhea, and applied externally to cuts and sores; bark combined with other medicines for boils, ulcers, and hemorrhage of the lungs
18	Picea x lutzii (hybrid Sitka spruce)	pitch combined with animal oils used for tuberculosis; pitch also used for measles
19	Pinus contorta (lodgepole pine)	cambium taken as blood purifier and purgative; resinous wood shavings and young needles taken as purgative and diuretic for gonorrhea and tuberculosis; pitch combined with other medicines to treat boils and ulcers
23	Angelica genuflexa (kneeling angelica)	decoction of roots combined with highbush cranberry bark taken for headache and eye troubles
24	Heracleum lanatum (cow-parsnip)	pulverized roots applied to rheumatic and other swellings
28	Oplopanax horridus (devil's club)	considered to be one of the best of all remedies; decoction taken to help broken bones heal and as purgative in the treatment of gonorrhea
29	Achillea millefolium (yarrow)	decoction of aboveground portions used as sore throat medicine
36	Alnus tenuifolia (mountain alder) or possibly A. crispa (green or Sitka alder)	decoction of bark and roots used as cough medicine; infusion or decoction of crushed pistillate catkins used as physic or diuretic to treat gonorrhea
39	Sisymbrium officinale (hedge mustard)	pulverized plant applied to cuts
41	Lonicera involucrata (black twinberry)	infusion of inner bark, bark, or fresh juice of berries used as eyewash
42	Sambucus racemosa (coastal red elder)	infusion of bark of roots used as emetic and purgative
44	Viburnum edule (highbush cranberry)	decoction of bark and twigs used as cough medicine in treatment of tuberculosis; docotion of all parts used as physic for any sickness, or with devil's-club as a diuretic and hernia; decoction combined with kneeling angelica roots used for unspecified medicine
48	Sedum lanceolatum (lance-leaved stonecrop)	pulverized stems applied to cuts
49	Shepherdia canadensis (soopolallie)	decoction of roots combined with hybrid Sitka spruce twigs used as rheumatism medicine; decoction of bark used as cough medicine
51	Ledum groenlandicum (Labrador tea)	decoction of leaves used as diuretic

61	?Ribes lacustre (black gooseberry)	decoction of bark used for unspecified ailments
65	Nuphar polysepalum (yellow pond-lily)	infusion of scrapings of toasted rhizome taken for hemorrhage of the lungs and as male contraceptive
69	Anemone multifida (cut-leaved anemone)	plant said to have been eaten, or decoction taken internally, in sweat bath as rheumatism treatment
73	Thalictrum occidentale (western meadowrue)	juice of chewed root swallowed for headache, eye trouble, sore legs as expectorant and to improve blood circulation
80	Malus fusca (Pacific crab apple)	juice of peeled trunk used as eye medicine; decoction of bark scrapings taken internally as laxative and diuretic for tuberculosis and rheumatism
88	Sorbus sitchensis (Sitka mountain-ash) and/or S. scopulina (western mountain-ash)	raw fruits eaten as strong physic
91	Populus tremuloides (trembling aspen)	bark of roots chewed or pulverized and applied to cuts; decoction of bark of trunk taken as purgative
95	Castilleja miniata (common red paintbrush)	decoction of entire plant taken internally as purgative and diuretic for nose bleed, bleeding, stiff lungs, bad eyes, and lame back possibly caused by kidney trouble
97	Urtica dioica (stinging nettle)	decoction of stalks and roots drunk for lungs or bladder trouble, spitting of blood, and for any illness
99	?Calla palustris (wild calla)	decoction of roots taken internally for blindness, hemorrhage of the lungs and from the mouth, for shortness of breath, and influenza
105	Smilacina racemosa (false Solomon's-seal)	decoction of roots used as purgative for sore back, kidney ailments, and rheumatism; roots pulverized and applied to cuts
108	Veratrum viride (Indian hellebore)	pulverized root applied to boils and to treat poisoning of the blood or combined with other medicines to treat boils, ulcers, hemorrhage of the lungs, and rheumatism; smoke of the roots inhaled for bad dreams, influenza, and rheumatism; leaves used in sweat bath for rheumatism

Table 5. Summary of botanical species with ritual or spiritual roles among the Gitksan.		
Entry Number	Species	Ritual/Spiritual Role(s)
16	Thuja plicata (western redcedar)	inner bark used for head and neck rings used in ceremonial applications
23	Angelica genuflexa (kneeling angelica)	roots used by gamblers for luck
30	Anaphalis margaritacea (pearly everlasting)	used as coffin and grave decoration

Table 6. Summary of botanical species with mythological roles among the Gitksan.		
Entry Number	Species	Mythological Role(s)
66	Epilobium angustifolium (fireweed)	basis of phratry name and crest symbol
94	Geocaulon lividum (bastard toad-flax)	regarded in story as berry of 'Wii Get/'Wii Gat
101	Clintonia uniflora (queen's cup)	said in story to be spoon of 'Wii Get/'Wii Gat

Table 7. Summary of botanical species with miscellaneous cultural roles among the Gitksan.

Entry Number	Species	Miscellaneous Role(s)
2	Inonotus obliquus (cinder conk)	associated with traditional belief regarding future of children
4	?Cetraria pinastri (moonshine cetraria)	traditional association with plant species: Pinus contorta (lodgepole pine)
5	Lobaria spp., e.g., L. pulmonaria (lung lichen)	traditional association with animal species: various species of frogs and/or toads
12	?Lycopodium complanatum (ground-cedar)	traditional association with animal species: Lontra canadensis (river otter)
13	?Lycopodium dendroideum (ground-pine)	traditional association with animal species: Lontra canadensis (river otter)
22	Acer glabrum (Douglas or Rocky Mountain maple)	seeds shaken as toy rattle by children
23	Angelica genuflexa (kneeling angelica)	hollow stems used as toy blow-gun; traditional association with plant species: ?Delphinium glaucum, (tall larkspur); Heracleum lanatum, (cow-parsnip); Thalictrum occidentale, (western meadowrue); traditional association with animal species: various species of frogs and/or toads
24	Heracleum lanatum (cow-parsnip)	hollow stems used as toy blow-gun; traditional association with plant species: Angelica genuflexa, (kneeling angelica); ?Delphinium glaucum, (tall larkspur); Thalictrum occidentale, (western meadowrue)
25	Osmorhiza chilensis (mountain sweet-cicely)	traditional association with plant species: Apocynum androsaemifolium, (spreading dogbane)
27	Aralia nudicaulis (wild sarsaparilla)	traditional association with animal species: Ursus americanus (black bear)
38	Corylus cornuta (beaked hazelnut)	roots bent for use as hockey stick in ground hockey
45	Pachistima myrsinites (falsebox)	traditional association with plant species: Arctostaphylos uva-ursi (kinnikinnick);
47	Cornus stolonifera (red-osier dogwood)	wood used for toy bows
68	Actaea rubra (baneberry)	traditional association with animal species: Ursus americanus (black bear)
71	?Delphinium glaucum (tall larkspur)	traditional association with plant species: Angelica genuflexa, (kneeling angelica); Heracleum lanatum, (cow-parsnip); Thalictrum occidentale, (western meadowrue); traditional association with animal species: various species of frogs and/or toads
72	Ranunculus abortivus (kidney-leaved buttercup)	traditional association with animal species: Lontra canadensis (river otter)
73	Thalictrum occidentale (western meadowrue)	traditional association with plant species: Angelica genuflexa, (kneeling angelica); ?Delphinium glaucum, (tall larkspur); Heracleum lanatum, (cow-parsnip); traditional association with animal species: various species of frogs and/or toads
74	Agrimonia striata (grooved agrimony)	traditional association with plant species: Epilobium angustifolium (fireweed)

76	Aruncus dioicus (goatsbeard)	traditional association with plant species: Apocynum androsaemifolium (spreading dogbane)
78	Dryas drummondii (yellow mountain-avens)	traditional association with plant species: Betula papyrifera (paper birch)
82	Prunus virginiana (choke cherry)	pits used as shot in toy blow-guns
85	Rubus parviflorus (thimbleberry)	leaves tied into ball used in play; folded leaves bitten to form designs
86	Rubus pubescens (trailing raspberry) or possibly R. pedatus (five-leaved creeping raspberry)	traditional association with animal species: various species of frogs and/or toads
92	Salix sitchensis (Sitka willow)	traditional association with animal species: Castor canadensis (beaver)

Table 8. Summary of botanical species recognized and named among the Gitksan but which lack cultural roles.	
Entry Number	Species
1	unidentified fungus
3	unidentified fungus on hemlock
11	Equisetum hyemale (scouring-rush)
31	Artemisia michauxiana (Michaux's mugwort)
32	Aster conspicuus (showy aster)
33	Lactuca biennis (tall blue lettuce)
34	Sonchus arvensis (sow-thistle)
40	Campanula rotundifolia (common harebell)
52	Menziesia ferruginea (false azalea)
59	Geranium richardsonii (white Geranium)
63	Mentha arvensis (field mint)
64	Prunella vulgaris (self-heal)
89	Spiraea douglasii (hardhack)
102	Disporum hookeri (Hooker's fairybells)
105	?Disporum trachycarpum (rough-fruited fairybells)
106	Smilacina stellata (star-flowered false Solomon's-seal)
107	Streptopus amplexifolius (clasping twistedstalk)
109	Platanthera obtusata (one-leaved rein orchid)
112	unidentified plant

In terms of numbers of culturally significant species, the plant families most important to the Gitksan are the Rosaceae, Liliaceae and Ericaceae while several additional angiosperm families have from one to a few species with cultural roles. The gymnosperms are the next most well represented botanical group in terms of species with cultural roles. The lower plants and fungi are only minimally represented in this regard.

Based on the data reported by Smith, it is clear that plants and fungi were most important in traditional Gitksan culture as sources of foods, various useful materials and medicines. The fruits, roots or other underground organs, and cambium of several species of pteridophytes, gymnosperms and angiosperms provided important food materials that were either eaten fresh in season or were preserved for later use throughout the winter months. At least one plant—Ledum groenlandicum (Labrador tea)—provided leaves used for a beverage tea while other plants produced leaves useful in the preparation or preservation of various food products. Wood, bark and fibres were among the most frequently and intensively used of plant-based technological materials, having a variety of roles in the construction of houses, canoes, and a variety of implements used in the procurement of primarily animal-based food resources. Various types of bark, pitch, roots and other plant materials were administered as decoctions, infusions or in other manners for their purgative, laxative, diuretic, emetic, expectorant, analgesic, antiseptic and other properties. With these medicines, the Gitksan treated a variety of diseases, including tuberculosis, gonorrhea and measles, as well as numerous other conditions including respiratory ailments, arthritis and rheumatism, renal difficulties, sore eyes, boils and other types of infection, headaches, bleeding noses, cuts and abrasions and broken bones. Additional interesting medicinal applications include the report of Nuphar polysepalum (yellow pond-lily) used as a male contraceptive and Veratrum viride (Indian hellebore) used to eliminate bad dreams.

While the majority of Gitksan ethnobotanical data that Smith recorded and presented involve primarily utilitarian applications, Smith did indicate that a few plants were important for their ritual or talismanic roles, or had interesting traditional beliefs or stories associated with them. For example, Angelica genuflexa (kneeling angelica) has been used in Gitksan hunting rituals and Anaphalis margaritacea (pearly everlasting) has been used for coffin and grave decoration. A type of shelf fungus (?Inonotus obliquus) was involved with traditional beliefs about how children may behave in their adulthood. Only a few plants are associated with mythology. Epilobium angustifolium (fireweed) is the basis for a Gitksan phratry name and crest symbol. Geocaulon lividum (bastard toad-flax) and Clintonia uniflora (Queen's cup) were both associated in stories with 'Wii Get (or 'Wii Gat)—the mythical personification of raven.

Smith underemphasized such cultural roles for botanical species and it is likely that further research on Gitksan ethnobotany would result in the documentation of more of this kind of information.

Several of the species listed in Smith's report seem only to have been recognized and named in Gitksan terms. These species lack cultural roles or are considered exceptional only because their names seem to indicate traditional associations between the plants they label and animal or other plant species.

Two plants are associated with black bear: Aralia nudicaulis (wild sarsaparilla) and Actaea rubra (baneberry). Their Gitksan names mean 'berry of black bear' and they are considered to be food only for bears. Several other botanical species are associated with frogs or toads and in some cases considered to be food for these creatures, viz., ?Lobaria pulmonaria (lung lichen), Angelica genuflexa (kneeling angelica), Delphinium glaucum (tall larkspur), Rubus pubescens (trailing raspberry) and Thalictrum occidentale (western meadowrue). One type of stonecrop (Sedum lanceolatum) is associated with raven, possibly to differentiate it from the edible spreading stonecrop (S. divergens). Still other plants (Lycopodium spp. and Ranunculus abortivus, clubmosses and kidney-leaved buttercup) were associated with river otters. Comparable data from other Tsimshianic languages and the North Wakashan languages suggest that among many First Nations groups of the Central Coast and North Coast regions recognized discrete conceptual classes of "black bear berries," "crow berries" or "frog berries"—plants that lack usefulness to humans but which produce berries eaten by, or thought to be eaten by, various animals (Compton 1993).

A total of 10 plants are associated with Gitksan names that suggest their similarity to other plants. For example, Osmorhiza chilensis (mountain sweet-cicely) is said to be named for its perceived similarity to Apocynum androsaemifolium (spreading dogbane). Campanula rotundifolia (common harebell) is named for its resemblance to an unidentified type of grass. Pachistima myrsinites (falsebox) is named for the fact that it is "nearly" the same as—or resembles—Arctostaphylos uva-ursi (kinnikinnick). However, many of the Gitksan terms or phrases associated with these species may actually represent descriptive nonce forms, rather than actual botanical names. Still other species (i.e., Aster conspicuus or showy aster, Lactuca biennis or tall blue lettuce, Mentha arvensis or field mint, and Sonchus arvensis or sow-thistle) are associated with descriptive terms that refer to their odour or habitat preference and may or may not represent valid Gitksan taxonomic labels. Further research may help either to validate these terms as legitimate conventional plant names or to reject them as nonce forms.[182]

The linguistic features of several of the Gitksan botanical terms recorded by Smith are indicative of some basic features of Gitksan botanical classification. Although no single, all-inclusive "plant" term exists among Smith's data or seems evident within the results of later workers (e.g, Hindle and Rigsby 1973), their are Gitksan terms that imply Gitksan recognition of several general plant categories.

In particular, the Gitksan linguistic element sgan is incorporated in 32 plant names recorded by Smith. This element and its variant sk'an correspond to Southern Tsimshian sxán- (literally, 'support-for...'), a prefix that is incorporated in Southern Tsimshian plant names and which always precedes nouns that label an important or otherwise relevant portion or other feature of a plant. As is the case with Southern Tsimshian, the application of Gitksan sgan seems limited to trees, shrubs and herbs of particular cultural significance or which posess unusual or distinctive attributes. Further, according to Smith's data, only names for gymnosperm or angiosperm species incorporate the Gitksan element, sgan. Smith, perhaps at the suggestion of the Gitksan consultants he worked with, glossed this element as it occured in plant names as "plant"—an indication that the inclusion of this element in a plant name implied reference to the entire plant, rather than just to the fruit or other important plant part.[183]

The prefix am-, meaning 'good for-,' also is relevant to a discussion of Gitksan plant names. It occurs in five plant terms documented by Smith—those for Thuja plicata (western redcedar), Alnus rubra (red alder), Betula papyrifera (paper birch) Populus balsamifera (black cottonwood) and Populus tremuloides (trembling aspen)—as well as in at least three others documented by later workers, i.e., for Abies amabilis (amabilis fir), Tsuga heterophylla (western hemlock) and an unidentified type of willow.[184] Its application identifies plant species that are considered to produce "good" or useful materials (e.g., inner redcedar bark), to be "good for" use in particular applications (e.g., black cottonwood wood for canoes) or to be

---

[182]These forms have not been attested by Gitksan speakers other than those consulted by Smith seven decades ago and these and other terms have simply been retranscribed or reconstituted from Smith's forms. For these reasons—and because comparable terms seem to be lacking from the other documented Tsimshianic botanical lexicons—these terms should be regarded as in need of verification with modern speakers.

[183]Although Smith documented several plant terms with sgan, he seems not to have systematically searched for all appropriate applications of this element with respect to plant names. Further exploration of the application of this element is required to fully characterize this linguistic element and its role in Gitksan botanical nomenclature and classification.

[184]These terms are amhoo'oxs, amgiikw and am ẁaasan, respectively (Hindle and Rigsby 1973).

as "good" as or better than some other species in one or another respect (i.e., trembling aspen which is associated with Douglas maple).[185]

Berries and other fruits are of considerable traditional and contemporary significance to the Gitksan; a fact implied by the ways in which several fruit-bearing plants are named in the Gitksan language. The general term that refers to any type of berry or other fruit—maa'y—also labels one particularly notable, or prototypical, fruit-bearing plant tentatively identified as Vaccinium membranaceum (black huckleberry).[186] Several other species and the fruits they produce are given names that incorporate another linguistic element: mii and the dialectal variant 'mii.[187] These include O. oxycoccus (bog cranberry), V. caespitosum (dwarf blueberry), Ribes hudsonianum (northern blackcurrant), Fragaria virginiana (wild strawberry), Rubus pubescens (trailing raspberry) and R. spectabilis (salmonberry).[188] Another species—Geocaulon lividum (bastard toad-flax)—is referred to as a 'berry' of 'Wii Get (or 'Wii Gat).[189]

The preceding observations and comments are by no means exhaustive and they are meant only to highlight some of the more unique and interesting topics represented in the data that Smith collected. In conclusion, it is clear that Smith's work on Gitksan ethnobotany represents the most comprehensive report to have been produced on the topic to date. At the same time, it may be regarded as evidence that further work on traditional Gitksan botanical knowledge, use and classification is merited, and could productively be pursued within the context of broader Tsimshianic studies. Such additional research would likely resolve many questions regarding the data documented and presented by Smith as well as contribute substantially to expanding the record of Gitksan interactions with plants and fungi.

---

[185]The reasons why these species should have names that incorporate the 'good for-' prefix is unclear but the same basic core tree species are similarly named in one or more other Tsimshianic languages.

[186]Another, perhaps more likely, candidate for an ideal, or prototypical 'berry' among the Gitksan is the fruit identified as sim maaẏ, literally, 'real/best berry.' This has also been identified as "huckleberry" (Hindle and Rigsby 1973:27) or as "huckleberry (Vaccinium parviflorum)" (People of 'Ksan 1980:126), an apparent reference to V. parvifolium Smith in Rees, red huckleberry. In contrast, the Coast Tsimshian and Nisga'a cognates of this term seem to refer to V. ovalifolium (Dunn 1978).

[187]Throughout the text of this document only the form mii has been used in the listing of Gitksan plant names with the understanding that variant forms incorporating 'mii may also occur.

[188]Salmonberry is curiously known in Gitksan, literally as 'berry maple.' The reason for this association, or the association of Douglas maple with trembling aspen, known as 'good for-maple,' is unexplained (cf. Compton 1993).

[189]Linguistic opinion is divided over the etymology of the linguistic element mii or 'mii, particularly with respect to its relationship to the word maa'y, referring to any berry or fruit.

References

Alcock, F.J. 1956. Annual Report of the National Museum of Canada for the Fiscal Year 1954-55. Bulletin No. 142. National Museum of Canada, Ottawa, Ontario.

Anonymous. 1926. Guards Formed at Kitwanga Village to Protect Totem Poles. Resources 4(11):15.

Bolton, L.L. 1925. 11 May 1925 letter to H.I. Smith. E. Sapir Correspondence (to H.I. Smith, 1921-1925). Canadian Museum of Civilization, Hull, Québec.

Cannings, Robert A. and Andrew P Harcombe (eds.) 1990. The Vertebrates of British Columbia: Scientific and Common Names. Royal British Columbia Museum Heritage Record No. 20; Wildlife Report No. R-24. Ministry of Municipal Affairs, Recreation and Culture and Ministry of Environment. Victoria, British Columbia.

Collins, W.H. 1928a. General Activities of the Museum. National Museum of Canada Bulletin No. 50, Annual Report for 1926 (pp. 1-31). King's Printer, Ottawa, Ontario.

————— 1929a. General Activities of the Museum. National Museum of Canada, Bulletin No. 56, Annual Report for 1927 (pp. 1-35). King's Printer, Ottawa, Ontario.

————— 1929b. General Activities of the Museum. National Museum of Canada, Bulletin No. 62, Annual Report for 1928 (pp. 1-21). King's Printer, Ottawa, Ontario.

Compton, Brian D. 1993. Upper North Wakashan and Southern Tsimshian Ethnobotany: The Knowledge and Usage of Plants and Fungi Among the Oweekeno, Hanaksiala (Kitlope and Kemano), Haisla (Kitamaat) and Kitasoo Peoples of the Central and North Coasts of British Columbia. Unpublished Ph.D. dissertation. Department of Botany. The University of British Columbia, Vancouver, British Columbia.

—————. 1995. "Ghost's Ears" (*Exobasidium* sp. affin. *vaccinii*) and Fool's Huckleberries (*Menziesia ferruginea* Smith): A Unique Report of Mycophagy on the Central Coast of British Columbia. Journal of Ethnobiology 15(1):89-98

Douglas, George W., Gerald B. Straley and Del Meidinger. 1989. The Vascular Plants of British Columbia, Part 1. - Gymnosperms and Dicotyledons (Aceraceae through Cucurbitaceae). Research Branch Special Report Series 1. Ministry of Forests, Victoria, British Columbia.

—————. 1990. The Vascular Plants of British Columbia, Part 2. - Dicotyledons (Diapensiaceae through Portulacaceae). Research Branch Special Report Series 2. Ministry of Forests, Victoria, British Columbia.

—————. 1991. The Vascular Plants of British Columbia, Part 3. - Dicotyledons (Primulaceae through Zygophyllaceae) and Pteridophytes. Research Branch Special Report Series 3. Ministry of Forests, Victoria, British Columbia.

———. 1994. The Vascular Plants of British Columbia, Part 4. - Monocotyledons. Research Branch Special Report Series 4. Ministry of Forests, Victoria, British Columbia.

Guédon, Marie-Françoise. 1973. List of plants collected around Hazelton, B.C. in June 1973 with preliminary information (available Gitksan name and medicinal uses). File VII-C-22M (B117 F5). Canadian Museum of Civilization, Hull, Québec.

Gottesfeld, Leslie M. Johnson. 1992. Use of Cinder Conk (Inonotus obliquus) by the Gitksan of Northwestern British Columbia, Canada. Journal of Ethnobiology 12(1):153-156.

___ and Beverley Anderson. 1988. Gitksan Traditional Medicine: Herbs and Healing. Journal of Ethnobiology 8(1):13-33.

Henry, J.K. 1915. Flora of Southern British Columbia and Vancouver Island. W.J. Gage & Co., Ltd., Toronto, Ontario.

Hindle, Lonnie and Bruce Rigsby. 1973. A Short Practical Dictionary of the Gitksan Language. Northwest Anthropological Research Notes 7(1):1-60.

Hitchcock, C. Leo and Arthur Cronquist. 1973. Flora of the Pacific Northwest. University of Washington Press, Seattle, Washington.

Krajina, V.J., K. Klinka and J. Worrall. 1982. Distribution and Ecological Characteristics of Trees and Shrubs of British Columbia. Faculty of Forestry, The University of British Columbia, Vancouver, British Columbia.

MacKinnon, Andy, Jim Pojar and Ray Coupé (eds.). 1992. Plants of Northern British Columbia. Lone Pine Publishing, Edmonton, Alberta.

Nater, H.F. 1990. A Concise Nuxalk-English Dictionary. Canadian Ethnology Service Mercury Series Paper 115. Canadian Museum of Civilization, Ottawa, Ontario.

People of 'Ksan, The. 1980. Gathering What the Great Nature Provided: Food Traditions of the Gitksan. Douglas and McIntyre, Vancouver, British Columbia.

Rigsby, Bruce and John Ingram. 1990. Obstruent Voicing and Glottalic Obstruents in Gitksan. International Journal of American Linguistics 56(2):251-263.

Sapir, E. 1925. 22 April 1925 letter to L.L. Bolton. E. Sapir Correspondence (to H.I. Smith, 1921-1925). Canadian Museum of Civilization, Hull, Québec.

Smith, Harlan I. 1913. 6 November 1913 letter to E. Sapir. E. Sapir Correspondence (to H.I. Smith, 1911-1920). Canadian Museum of Civilization, Hull, Québec.

———. 1920. 9 July 1920 letter to E. Sapir. E. Sapir Correspondence (to H.I. Smith, 1911-1920). Canadian Museum of Civilization, Hull, Québec.

———. 1920-1923a. The Material Culture of the Carrier Indians of British Columbia, Part I, 1-11: Introduction to Food Starvation. Microfiche VI-B-32M (B88 F1). Canadian Museum of Civilization, Hull, Québec.

———. 1920-1923b. The Material Culture of the Carrier Indians of British Columbia, Part II: 12-58-1/2, Securing Food to Dwellings. Microfiche VI-B-32M (B88 F2). Canadian Museum of Civilization, Hull, Québec.

———. 1920-1923c. The Uses of Plants by the Bella Coola Indians of British Columbia, Volume I, to 95. Microfiche VII-D-9M (B83 F3). Canadian Museum of Civilization, Hull, Québec.

———. 1920-1923d. The Uses of Plants by the Bella Coola Indians of British Columbia, Volume II, 101-200. Microfiche VII-D-9M (B83 F4). Canadian Museum of Civilization, Hull, Québec.

———. 1920-1923e. The Uses of Plants by the Bella Coola Indians of British Columbia, Volume III, 201 to End and Bibliography. Microfiche VII-D-9M (B83 F1). Canadian Museum of Civilization, Hull, Québec.

———. 1920-1923f. The Uses of Plants by the Carrier Indians of British Columbia, Volume I, to 199. Microfiche VI-B-21M (B86 F3). Canadian Museum of Civilization, Hull, Québec.

———. 1920-1923g. The Uses of Plants by the Carrier Indians of British Columbia, Volume II, 200 to End. Microfiche VI-B-21M (B86 F4). Canadian Museum of Civilization, Hull, Québec.

———. 1922. 21 April 1922 letter to E. Sapir. E. Sapir Correspondence (to H.I. Smith, 1921-1925). Canadian Museum of Civilization, Hull, Québec.

———. 1925. The Conservation of Beaver by an Indian. Science 62:461.

———. 1925-1926. Department of Indian Affairs, Totem Pole Preservation at Kitwanga, B.C. 1925 and 1926. Blueprints in the Harlan I. Smith Collection. File 1192.4C B91 F1 VII-C-84M. Canadian Museum of Civilization, Hull, Québec.

———. 1925-1927. Ethno-botany of the Gitksan Indians of British Columbia. Microfiche VII-C-81M (B90 F1). Canadian Museum of Civilization, Hull, Québec.

———. 1926. Guide to the Totem Poles of Kitwanga, British Columbia on the Triangle Route of the Canadian National Railways. Microfiche VII-C-82M (B90 F2). Victoria Memorial Museum, Ottawa, Ontario.

———. 1927. Handbook of the Kitwanga Garden of Native Plants Under the Charge of the Totem Pole Indians, Kitwanga, British Columbia. National Museum of Canada and Department of Indian Affairs, Ottawa, Ontario. (Note: This document is contained along with the undated manuscript entitled, "The Kitwanga Garden of Native Plants, Kitwanga, British Columbia" by Harlan I. Smith on Microfiche VII-C-83M [B90 F3, Canadian Museum of Civilization, Hull, Québec.)

———. (no date) The Kitwanga Garden of Native Plants, Kitwanga, British Columbia. Microfiche VII-C-83M (B90 F3). Canadian Museum of Civilization, Hull, Québec. (Note: This document is contained along with the 1927 document entitled, "Handbook of the Kitwanga Garden of

Native Plants" by Harlan I. Smith on Microfiche VII-C-83M [B90 F3], Canadian Museum of Civilization, Hull, Québec.)

———. 1928. Restoration of Totem-Poles in British Columbia. National Museum of Canada Bulletin No. 50, Annual Report for 1926 (pp. 81-83). King's Printer, Ottawa, Ontario.

———. 1929. Materia Medica of the Bella Coola and Neighbouring Tribes of British Columbia. National Museum of Canada Bulletin No. 56, Annual Report for 1927 (pp. 47-68). King's Printer, Ottawa, Ontario.

———. 1936. 2 November 1936 letter to Publicity Department, Canadian National Railways, Montreal, Québec. H.I. Smith Correspondence, File I-A-242M (B216) FC. Canadian Museum of Civilization, Hull, Québec.

Steedman, Elsie Viault, ed. 1930. Ethnobotany of the Thompson Indians of British Columbia based on field notes by James A. Teit. Pp. 441-522 in Forty-fifth Annual Report of the Bureau of American Ethnology, 1927-1928. Washington, D.C.

Taylor, Roy L. and Bruce MacBryde. 1977. Vascular Plants of British Columbia. The University of British Columbia Press, Vancouver, British Columbia.

Turner, Nancy J. 1973. The Ethnobotany of the Bella Coola Indians of British Columbia. Syesis 6:193-220.

———. 1988. "The Importance of a Rose": Evaluating the Cultural Significance of Plants in Thompson and Lillooet Interior Salish. American Anthropologist 90(2):272-290.

———, Leslie M. Johnson Gottesfeld, Harriet V. Kuhnlein, Adolf Ceska. 1992. Edible Wood Fern Rootstocks of Western North America: Solving an Ethnobotanical Puzzle. Journal of Ethnobiology 12(1):1-34.

Wintemberg, W.J. 1940. Harlan Ingersoll Smith. American Antiquity 6(1):63-64.

Appendix 1. Writing the Gitksan Language.

We employ the practical alphabet and writing system here that Bruce Rigsby and Lonnie Hindle developed in the early 1970s and used in their A Practical Dictionary of the Gitksan Language (Hindle and Rigsby 1973, see especially pp 5-7). Since then, the alphabet and writing system have gained wider acceptance among Gitksan people, and they are used in the Gitksan language programs taught in schools of the region. The Nisgha and Tsimshian practical alphabets and writing systems are similar, but not identical.

The Gitksan alphabet uses these letters and letter-combinations to represent recurrent sounds and sound-sequences. In alphabetical order, these are: a, aa, b, d, e, ee, g, g̲, gw, h, hl, i, ii, j, k, k', k̲, k̲', kw, kw', l, 'l, m, 'm, n, 'n, o, oo, p, p', s, t, t', tl', ts, ts', u, uu, w, 'w, x, x̲, xw, y, 'y, and ' (glottal stop).

Single vowels are short, double vowels are long. Consonant sounds generally are represented like their closest English equivalents, e.g., b, d, g, etc. g̲ is a voiced uvular stop. gw is a voiced labiovelar stop. hl is a voiceless lateral fricative. j is a voiced rill affricate, similar to English dz and sometimes j. k' is a glottalized palatal stop. k̲ is a voiceless uvular stop. k̲' is a glottalized uvular stop. kw is a voiceless labiovelar stop. kw' is a glottalized labiovelar stop. 'l is a preglottalized lateral vocoid. 'm is a preglottalized bilabial nasal. 'n is a preglottalized alveolar nasal. p' is a glottalized bilabial stop. s is often like English s, but sometimes approaches English sh. t' is a glottalized alveolar stop. tl' is a glottalized lateral affricate. ts is a voiceless rill affricate, and sometimes it represents a cluster of t followed by s. 'w is a preglottalized labial glide. x̲ is a voiceless uvular fricative. xw is a voiceless labiovelar fricative. 'y is a preglottalized palatal glide. ' is the glottal stop, except that - represents the glottal stop following a stop consonant to distinguish it from the homophonous glottalized stop.

The practical alphabet and writing system indicate a level of representation that corresponds to what linguists would call broad phonetic transcription or systematic phonetic representation, but it does not include free variation. Gitksan phonemic or phonological representation is more abstract than this. A technical discussion of major features of Gitksan phonology is found in Rigsby and Ingram (1990).

Appendix 2. Botanical species collected or observed by H.I. Smith in 1925 and 1926.[190]

Part 1: 1925 Collections

Smith: 169 (1) Prunus demissa Nutt. (Choke Cherry) (from Kitwanga, B.C., 1925)
Douglas et al.: Prunus virginiana L. ssp. demissa Taylor & MacBryde (Choke Cherry)

Smith: 281 (1A) + (1B) Viburnum pauciflorum Raf. (Squashberry, High Bush Cranberry) (from Kitwanga, B.C., August 17, 1295)
Douglas et al.: Viburnum edule (Michx.) Raf. (Highbush Cranberry)

Smith: 280 (2) Symphoricarpus racemosus Michx. (Wax Berry, Snowberry) (from Kitwanga, B.C., August 17, 1925)
Douglas et al.: Symphoricarpos albus (L.) Blake (Common Snowberry)

Smith: (3) No specimen. See 14 [Arctostaphylos uva-ursi]
Douglas et al.: Arctostaphylos uva-ursi (L.) Spreng. (Kinnikinnick, Common Bearberry, Mealberry, or Sandberry)

Smith: (4) No specimen. See 15 [lichen from scrub pine]

Smith: 106 (5) Comandra livida Richards (Bastard Toad-flax) (from first terrace, south of Skeena River, near Kitwanga, B.C., August 23, 1925.)
Douglas et al.: Geocaulon lividum (Richards.) Fern. (Bastard Toad-flax, or Northern Comandra)

Smith: 234 (6) Monotropa uniflora L. (Indian Pipe) (from lower terrace near falls, south side of Skeena River, about two miles east of Kitwanga, B.C., August 23, 1925)
Douglas et al.: Monotropa uniflora L. (Indian-pipe)

Smith: 104 (7) Alnus tenuifolia Nutt. (Mountain Alder) (from Kitwanga, B.C., August 23, 1925)
Douglas et al.: Alnus tenuifolia Nutt. (Mountain Alder)

---

[190]Smith's original title for this section of his original document was: "List of plants collected by Harlan I. Smith beginning in 1925 near Kitwanga B.C. as identified by Dr. M.O. Malte of which Gitksan Indians of Kitwanga and vicinity gave ethno-botanical information." The order of arrangement and numbering of taxa is as originally presented by Smith. This appendix contains the botanical terminology and ancillary comments provided by Smith along with synonymous botanical nomenclature from Douglas et al. (1989, 1990, 1991, 1994). Note that editorial comments in this section appear in square brackets.

Smith: 226 (8) <u>Cornus stolonifera</u> Michx. (Red-osier Dogwood, Red Willow) (Kinnikinnik)[191] (from Kitwanga, B.C., 1925)

Douglas et al.: <u>Cornus stolonifera</u> Michx. (Red-osier Dogwood)

Smith: 279 (9) <u>Lonicera involucrata</u> Banks (Black Twin-berry) (from Kitwanga, B.C., August 23, 1925)

Douglas et al.: <u>Lonicera involucrata</u> (Richards.) Banks ex Spreng. (Black Twinberry, or Bearberry Honeysuckle)

Smith: 183 (10) <u>Amelanchier florida</u> Lindl. (Juneberry, Saskatoon) (Kitwanga, B.C., August 23, 1925)

Douglas et al.: <u>Amelanchier alnifolia</u> (Nutt.) Nutt. (saskatoon)

Smith: 226 (11) <u>Cornus canadensis</u> L. (Bunchberry, Dwarf Dogwood) (from Kitwanga, B.C., August 23, 1925)

Douglas et al.: <u>Cornus canadensis</u> L. (Bunchberry, or Canadian Bunchberry)

Smith: 170 (12) <u>Prunus emarginata</u> Dougl. (Wild Cherry) (from Kitwanga, B.C., August 23, 1925)

Douglas et al.: <u>Prunus emarginata</u> (Dougl.) Walp.(Bitter Cherry)

Smith: 81 (13) <u>Smilacina racemosa</u> L. (False Solomon's Seal) (from Kitwanga, B.C., August 23, 1925)

Douglas et al.: <u>Smilacina racemosa</u> (L.) Desf. var. <u>amplexicaulis</u> (Nutt. ex Baker) S. Wats. (False Solomon's-seal)

Smith: 232 (14) <u>Arctostaphylos uva-ursi</u> Spreng. (Kinnikinnik, Bear berry) (from Kitwanga, B.C., August 23, 1925)

Douglas et al.: <u>Arctostaphylos uva-ursi</u> (L.) Spreng. (Kinnikinnick, Common Bearberry, Mealberry, or Sandberry)

(15) (See 4) (Lichen from Scrub Pine) (See 21) (from Kitwanga, B.C., August 23, 1925)

---

[191]Here, Smith associated the common name "kinnikinnik" with <u>C. stolonifera</u>, rather than with, <u>Arctostaphylos uva-ursi</u>, the species typically meant by this common name in British Columbia.

Smith: 182 (16) Crataegus brevispina (Dougl.) Hell. (Black Hawthorn, Haw) (from bottom lands, Kitwanga, B.C., Sept. 3, 1925)
Douglas et al.: Crataegus douglasii Lindl. (Black Hawthorn)

Smith: 183 (17) Pyrus diversifolia Bong. (Crab Apple) (from bottom lands, Kitwanga, B.C., Sept. 3, 1925)
Douglas et al.: Malus fusca (Raf.) Schneid. (Pacific Crab Apple)

Smith: 221 (18) Heracleum lanatum Michx. (Cow Parsnip) (from Kitwanga, B.C., Kitwankool road, 1925)
Douglas et al.: Heracleum lanatum Michx. (Cow-parsnip)

Smith: 210 (19) Shepherdia canadensis Nutt. (Soopolallie, Soapolallie) (Seen on a dry creek bed on the terrace on the south side of Skeena River opposite Kitwanga, B.C., August 20, 1925.)
Douglas et al.: Shepherdia canadensis (L.) Nutt. (Soopolallie, Soapberry, or Canadian Buffalo-berry)

Smith: 104 (20) Alnus rubra Bong. (Red Alder) (Seen at Kitwanga, B.C., August 9, 1925.)
Douglas et al.: Alnus rubra Bong. (Red Alder)

Smith: 170 (21) Rubus parviflorus Nutt. (Thimbleberry, Red Cap) (Seen on the gravel terraces near Kitwanga, B.C., August 20, 1925.)
Douglas et al.: Rubus parviflorus Nutt. (Thimbleberry)

Smith: 228 (22) Vaccinium Oxycoccus var. intermedium Gray (Cranberry)
Douglas et al.: Oxycoccus oxycoccus (L.) MacM. (Bog Cranberry)

Smith: 174 (23) Rosa sp. (Wild Rose) (Seen on gravel terrace near Kitwanga, B.C., September 16, 1925.)

Smith: 212 (24) Epilobium angustifolium L. (Fire-weed) (Seen on the sides of the Valley near Kitwanga, B.C., September 16, 1925.)
Douglas et al.: Epilobium angustifolium L. ssp. angustifolium and E. angustifolium L. ssp. circumvagum Mosquin (Fireweed)

Smith: 183 (25) <u>Pyrus</u> <u>sitchensis</u> (Roem) Piper or <u>P</u>. <u>occidentalis</u> Wats. (Mountain Ash)
Douglas et al.: <u>Sorbus</u> <u>sitchensis</u> Roemer (Sitka Mountain-ash)

Smith: 292 (26) <u>Aster</u> sp. (Blue Aster)

Smith: 307 (27) <u>Achillea</u> <u>lanulosa</u> Nutt., or <u>A</u>. <u>millefolium</u> L. (Yarrow)
Douglas et al.: <u>Achillea</u> <u>millefolium</u> L. (Yarrow)

Smith: 300 (28) <u>Solidago</u> sp. (Golden Rod)

Smith: 105 (29) <u>Urtica</u> <u>Lyallii</u> Wats. (Western Nettle) (Seen at Kitwanga, B.C., August 9, 1925.)
Douglas et al.: <u>Urtica</u> <u>dioica</u> L. ssp. <u>gracilis</u> (Ait.) Seland. (Stinging Nettle)

Part 2: 1926 Collections[192]

Smith: 30. <u>Lonicera</u> <u>involucrata</u> Banks
Douglas et al.: <u>Lonicera</u> <u>involucrata</u> (Richards.) Banks ex Spreng. (Black Twinberry, or Bearberry Honeysuckle)

Smith: 31. Lichen ["frog blanket," i.e., <u>Lobaria</u> spp.]

Smith: 32. <u>Fragaria</u> <u>glauca</u> (Wats.) Rydb.
Douglas et al.: <u>Fragaria</u> <u>virginiana</u> Duch. ssp. <u>glauca</u> (S. Wats.) Staudt (Wild Strawberry)

Smith: 33. <u>Sambucus</u> <u>racemosa</u> L.
Douglas et al.: <u>Sambucus</u> <u>racemosa</u> L. ssp. <u>pubens</u> (Michx.) House var. <u>arborescens</u> and possibly also <u>S</u>. <u>racemosa</u> L. ssp. <u>pubens</u> (Michx.) House var. <u>leucocarpa</u> (Coastal Red Elder or Elderberry)

Smith: 34a. <u>Equisetum</u> <u>arvense</u> L.
Douglas et al.: <u>Equisetum</u> <u>arvense</u> L. (Common or Field Horsetail)

Smith: 34b. <u>Equisetum</u> <u>hyemale</u> L.

---

[192] Smith indicated that he collected these specimens during 1926 and sent them for identification to Dr. Malte on 19 August and 3 November 1926.

Douglas et al.: Equisetum hyemale L. ssp. affine (Engelm.) Stone. (Scouring-rush)

Smith: 35. Acer glabrum Torr.
Douglas et al.: Acer glabrum Torr. var. douglasii (Hook.) Dippel (Douglas or Rocky Mountain Maple)

Smith: 36. Corylus rostrata Ait.
Douglas et al.: Corylus cornuta Marsh. var. cornuta (Beaked Hazelnut)

Smith: 37. Asplenium cyclosorum Rupr.
Douglas et al.: Athyrium filix-femina (L.) Roth subsp. cyclosorum (Rupr.) C. Christ. in Hult. (Lady Fern)

Smith: 38. Lathyrus ochroleucus Hook.
Douglas et al.: Lathyrus ochroleucus Hook. (Creamy or Cream-colored Peavine)

Smith: 39. Thalictrum occidentale Gray
Douglas et al.: Thalictrum occidentale A. Gray (Western Meadowrue)

Smith: 40. Aquilegia formosa Fischer
Douglas et al.: Aquilegia formosa Fisch. in DC. (Red or Sitka Columbine)

Smith: 41. Ribes Hudsonianum Richards
Douglas et al.: Ribes hudsonianum Richards. in Frank. (Northern Blackcurrant)

Smith: 42. Rubus pubescens Raf.
Douglas et al.: Rubus pubescens Raf. (Trailing Raspberry, or Dwarf Red Blackberry)

Smith: 43. Smilacina racemosa L.
Douglas et al.: Smilacina racemosa (L.) Desf. var. amplexicaulis (Nutt. ex Baker) S. Wats. (False Solomon's-seal)

Smith: 44. Streptopus amplexifolius DC.
Douglas et al.: Streptopus amplexifolius (L.) DC. in Lam. & DC. var. americanus Schult., S. amplexifolius (L.) DC. in Lam. & DC. var. chalazatus Fassett (Clasping Twistedstalk)

Smith: 45. Ranunculus abortivus L.

Douglas et al.: Ranunculus abortivus L. (Kidney-leaved Buttercup)

Smith: 46. Aralia nudicaulis L.

Douglas et al.: Aralia nudicaulis L. (Wild Sarsaparilla)

Smith: 47. Viola adunca Sm.

Douglas et al.: Viola adunca J.E. Smith in Rees (Early Blue Violet)

Smith: 48. Salix sp. (probably S. sitchensis Bong.)

Douglas et al.: Salix sitchensis Sanson ex Bong. (Sitka Willow)

Smith: 49. Viola canadensis L.

Douglas et al.: Viola canadensis L. ssp. rydbergii (Greene) House in Rydb. (Canada Violet)

Smith: 50. Sisymbrium incisum Eng.? (too young)

Douglas et al.: Sisymbrium officinale (L.) Scop. (Hedge Mustard)

Smith: 51. Calla palustris L.

Douglas et al.: Calla palustris L. (Wild Calla)

Smith: 52. Vaccinium caespitosum Michx.

Douglas et al.: Vaccinium caespitosum Michx. (Dwarf Blueberry or Bilberry)

Smith: 52A. Anemone globosa Nutt. (multifida Poir)

Douglas et al.: Anemone multifida Poir. (Cut-leaved or Pacific Anemone)

Smith: 52B. Antennaria rosea Greene

Douglas et al.: Antennaria microphylla Rydb. (Rosy Pussytoes)

Smith: 52C. Arabis hirsuta (L.) Scop.

Douglas et al.: Arabis hirsuta (L.) Scop. (Hairy Rockcress)

Smith: 52D. Crepis elegans Hook.

Douglas et al.: Crepis elegans Hook. (Elegant Hawksbeard)

Smith: 52E. Senecio balsamitae Muhl.
Douglas et al.: Senecio pauperculus Michx. (Canadian Butterweed)

Smith: 52F. Saxifraga tricuspidata Rottb.
Douglas et al.: Saxifraga tricuspidata Rottb. (Three-toothed Saxifrage)

Smith: 52G. Erigeron philadelphicus L.
Douglas et al.: Erigeron philadelphicus L. (Philadelphia Fleabane or Daisy)

Smith: 52H. Polemonium pulcherrimum Hook.
Douglas et al.: Polemonium pulcherrimumHook. ssp. pulcherrimum (Showy Jacob's-ladder)

Smith: 53. Sedum stenopetalum Pursh[193]
Douglas et al.: Sedum stenopetalum Pursh (Worm-leaved or Narrow-petaled Stonecrop)

Smith: 54. Mertensia paniculata Don (June 6)
Douglas et al.: Mertensia paniculata (W. Ait.) G. Don (Tall Bluebells, or Panicled Mertensia)

Smith: 55. Lupinus arcticus Wats. (June 8)
Douglas et al.: Lupinus arcticus S. Wats. (Arctic Lupine)

Smith: 56. Menyanthes trifoliata L.
Douglas et al.: Menyanthes trifoliata L. (Buckbean)

Smith: 57. Actaea eburnea Rydb.
Douglas et al.: Actaea rubra (Ait.) Willd. (Baneberry)

Smith: 57. Oxytropis monticola Gray
Douglas et al.: Oxytropis monticola A. Gray ssp. monticola (Mountain Locoweed)

Smith: 58. Linnaea borealis L. var. americana (Forbes) Rehder
Douglas et al.: Linnaea borealis L. (Twinflower)

Smith: 59. Collinsia parviflora Dougl.

---

[193] As stated earlier in the main text, Sedum stenopetalum is likely an erroneous identification where S. lanceolatum is the species more likely to have been collected by Smith.

Douglas et al.: <u>Collinsia parviflora</u> Doug. ex Lindl. (Small-flowered Blue-eyed Mary)

Smith: 60. <u>Gilia gracilis</u> Hook.
Douglas et al.: <u>Microsteris gracilis</u> (Hook.) Greene (Pink Microsteris)

Smith: 61. <u>Trientalis arctica</u> Fisch.
Douglas et al.: <u>Trientalis arctica</u> Fisch. ex Hook. (Northern Starflower)

Smith: 62. <u>Smilacina stellata</u> (L.) Desf.
Douglas et al.: <u>Smilacina stellata</u> (L.) Desf. (Star-flowered False Solomon's-seal)

Smith: 63. [no data provided by Smith, but this number is associated with <u>Juniperus</u> sp. in the main text]
Douglas et al.: <u>Juniperus communis</u> L. (Common or Ground Juniper)

Smith: 64. <u>Mentha canadensis</u> L. var. <u>glabrata</u> Benth.
Douglas et al.: <u>Mentha arvensis</u> L. (Field Mint)

Smith: 65. [no data provided by Smith, but this number seems to refer to <u>Elymus</u> sp.]
Douglas et al.: ?<u>Elymus glaucus</u> Buckl. (Blue Wildrye) or <u>E</u>. <u>trachycaulus</u> (Link) Gould in Shinners (Slender Wheatgrass)

Smith: 66. [no data provided by Smith]

Smith: 67. [no data provided by Smith, but this number seems to refer to <u>Rubus spectabilis</u>]
Douglas et al.: <u>Rubus spectabilis</u> Pursh (Salmonberry)

Smith: 68. [no data provided by Smith, but this number seems to refer to <u>Rubus idaeus</u> subsp. <u>melanolasius</u>]
Douglas et al.: <u>Rubus idaeus</u> L. ssp. <u>melanolasius</u> (Dieck) Focke (Red Raspberry, or American Red Raspberry)

Smith: 69. <u>Valeriana septentrionalis</u> Rydb.[194]
Douglas et al.: <u>Valeriana dioica</u> L. (Marsh Valerian)

---

[194]Smith noted that this specimen was probably a gift of John Laknitz.

Smith: 70. Potentilla viridescens Rydb.
Douglas et al.: Potentilla gracilis Dougl. (Graceful Cinquefoil)

Smith: 71. Galium boreale L.
Douglas et al.: Galium boreale L. (Northern Bedstraw)

Smith: 72. Senecio cymbalarioides Nutt.
Douglas et al.: Senecio cymbalarioides Buek non Nutt. (Alpine Meadow Butterweed)

Smith: 73. Rumex acetosella L.
Douglas et al.: Rumex acetosella L. (Sheep Sorrel)

Smith: 74. Erysimum cheiranthoides L.
Douglas et al.: Erysimum cheiranthoides L. (Wormseed Mustard)

Smith: 75. Actaea arguta Nutt.
Douglas et al.: Actaea rubra (Ait.) Willd. (Baneberry)

Smith: 76. Anaphalis margaritacea Benth. (Hazelton, July 18)
Douglas et al.: Anaphalis margaritacea (L.) Benth. & Hook. f. ex C.B. Clarke (Pearly Everlasting)

Smith: 77. Aster conspicuus Lindl. (Hazelton, July 18)
Douglas et al.: Aster conspicuus Lindl. in Hook. (Showy Aster)

Smith: 78. [no data provided by Smith, but this number seems to refer to Vaccinium caespitosum]
Douglas et al.: Vaccinium caespitosum Michx. (Dwarf Blueberry or Bilberry)

Smith: 79. Delphinium Brownii Rydb. (Hazelton, July 18)
Douglas et al.: ?Delphinium glaucum S. Wats. (Tall Larkspur)

Smith: 80. Aruncus sylvester Kost.
Douglas et al.: Aruncus dioicus (Walt.) Fern. (Goatsbeard)

Smith: 82. Prunella vulgaris L.
Douglas et al.: Prunella vulgaris L. ssp. lanceolata (Bart.) Hult. (Self-heal)

Smith: 83. Parnassia montanensis Rydb. & Fer.
Douglas et al.: Parnassia palustris L. (Northern Grass-of-Parnassus)

Smith: 84. Osmorrhiza divaricata Nutt.
Douglas et al.: Osmorhiza chilensis H. & A. (Mountain Sweet-cicely)

Smith: 85. Disporum oreganum B. & Hook.
Douglas et al.: Disporum hookeri (Torr.) Nichols. var. oreganum (S. Wats.) Q. Jones (Hooker's Fairybells)

Smith: 86. Smilacina racemosa L.
Douglas et al.: Smilacina racemosa (L.) Desf. var. amplexicaulis (Nutt. ex Baker) S. Wats. (False Solomon's-seal)

Smith: 87. Ribes Hudsonianum Richards
Douglas et al.: Ribes hudsonianum Richards. in Frank. (Northern Blackcurrant)

Smith: 88. [no data provided by Smith, but this number seems to refer to Fritillaria camschatcensis or Smilacina sp.]
Douglas et al.: Fritillaria camschatcensis (L.) Ker-Gawl. (Northern Rice-root, or Riceroot Fritillary)
Douglas et al.: Smilacina racemosa (L.) Desf. var. amplexicaulis (Nutt. ex Baker) S. Wats. (False Solomon's-seal)

Smith: 89. Lactuca spicata (Lam.) Hitchc.
Douglas et al.: Lactuca biennis (Moench) Fern. (Tall Blue Lettuce)

Smith: 90. Heuchera glabra Willd.
Douglas et al.: Heuchera glabra Willd. ex Roem. & Schult. (Smooth Alumroot)

Smith: 91. Galium triflorum Michx.
Douglas et al.: Galium triflorum Michx. (Sweet-scented Bedstraw)

Smith: 93. Castilleja miniata Dougl. (Kitwanga, June 21)

Douglas et al.: Castilleja miniata Dougl. ex Hook. (Scarlet or Common Red Paintbrush)

Smith: 94. Spiraea Douglasii Hook. var. Menziesii Presl. (Kitwanga, June 21)

Douglas et al.: Spiraea douglasii Hook. ssp. menziesii (Hook.) Calder & Taylor (Hardhack, Pink Spiraea or Menzies' Spirea)

Smith: 95. Galium boreale L. (Kitwanga, June 21)

Douglas et al.: Galium boreale L. (Northern Bedstraw)

Smith: 96. Gilia linearis Gray (Kitwanga, June 31)

Douglas et al.: ?Collomia linearis Nutt. (Narrow-leaved Collomia)

Smith: 97. Polygonum convolvulus L. (Kitwanga, June 21)

Douglas et al.: Polygonum convolvulus (L.) (Black Bindweed)

Smith: 98. Rhinanthus crista-galli L. (Kitwanga, June 21)

Douglas et al.: Rhinanthus minor L. (Yellow Rattle)

Smith: 99. Agrimonia gryposepala Wallr. (Kitwanga, June 21)

Douglas et al.: Agrimonia gryposepala Wallr. (Common Agrimony)

Smith: 100. Clintonia uniflora Kunth (Kitwanga, June 21)

Douglas et al.: Clintonia uniflora (Schult.) Kunth (Queen's Cup, or Blue-bead Clintonia)

Smith: 101. Prunus emarginata Dougl.

Douglas et al.: Prunus emarginata (Dougl.) Walp.(Bitter Cherry)

Smith: 102. Prunus demissa Nutt.

Douglas et al.: Prunus virginiana L. ssp. demissa Taylor & MacBryde, P. virginiana L. ssp. melanocarpa (Nels.) Taylor & MacBryde (Choke Cherry)

Smith: 103. Sedum stenopetalum Pursh. (June 9, Kitwanga)[195]

Douglas et al.: Sedum stenopetalum Pursh (Worm-leaved or Narrow-petaled Stonecrop)

---

[195] See the earlier footnote pertaining to this species.

Smith: 104. <u>Geranium</u> <u>erianthum</u> DC.
Douglas et al.: <u>Geranium</u> <u>erianthum</u> DC. (Northern Geranium or Crane's-bill)

Smith: 105. <u>Capsella</u> <u>bursa-pastoris</u> (L.) Medic. (June 8)
Douglas et al.: <u>Capsella</u> <u>bursa-pastoris</u> (L.) Medic. (Shepherd's purse)

Smith: 106. <u>Antennaria</u> <u>Howellii</u> Greene (June 8)
Douglas et al.: <u>Antennaria</u> <u>neglecta</u> Greene (Field Pussytoes)

Smith: 107. <u>Arabis</u> <u>hirsuta</u> (L.) Scop. (June 8)
Douglas et al.: <u>Arabis</u> <u>hirsuta</u> (L.) Scop. (Hairy Rockcress)

Smith: 108. <u>Arnica</u> <u>cordifolia</u> Hook. (June 8)
Douglas et al.: <u>Arnica</u> <u>cordifolia</u> Hook. (Heart-leaved Arnica)

Smith: 109. <u>Corydalis</u> <u>aurea</u> Willd. (June 8)
Douglas et al.: <u>Corydalis</u> <u>aurea</u> Willd. ssp. <u>aurea</u> (Golden Corydalis)

Smith: 110. <u>Actaea</u> <u>arguta</u> Nutt.? (June 8)
Douglas et al.: <u>Actaea</u> <u>rubra</u> (Ait.) Willd. (Baneberry)

Smith: 111. <u>Erigeron</u> <u>philadelphicus</u> L. (Hazelton, May 24)
Douglas et al.: <u>Erigeron</u> <u>philadelphicus</u> L. (Philadelphia Fleabane or Daisy)

Smith: 112. <u>Anemone</u> <u>globosa</u> Nutt. [syn.: <u>A</u>. <u>multifida</u> Poir.]
Douglas et al.: <u>Anemone</u> <u>multifida</u> Poir. (Cut-leaved or Pacific Anemone)

Smith: 113. <u>Draba</u> <u>lutea</u> Gilib.
Douglas et al.: ?<u>Draba</u> <u>aurea</u> Vahl in Horn. (Golden Draba or Whitlow-grass)

Smith: 114. <u>Angelica</u> <u>genuflexa</u> Nutt.
Douglas et al.: <u>Angelica</u> <u>genuflexa</u> Nutt. (Kneeling Angelica)

Smith: 115. <u>Lycopodium</u> <u>obscurum</u> L.
Douglas et al.: <u>Lycopodium</u> <u>dendroideum</u> Michx. (Ground-pine)

Smith: 116. Moss

Smith: 117. Cirsium (Carduus) undulatum (Nutt.) Spreng.
Douglas et al.: Cirsium undulatum (Nutt.) Spreng. (Wavy-leaved Thistle)

Smith: 118. Spiraea Douglasii Hook. var. Menziesii Presl. (late September 1926)
Douglas et al.: Spiraea douglasii Hook. ssp. menziesii (Hook.) Calder & Taylor (Hardhack, Pink Spiraea or Menzies' Spirea)

Smith: 119. Dryas Drummondii Rich. (Nash, September 6)
Douglas et al.: Dryas drummondii Richards. in Hook. var. drummondii, D. drummondii Richards. in Hook. var. tomentosa (Farr) Williams (Yellow Mountain-avens)

Smith: 120. Lesquerella Douglasii Wats. (Nash, September 6)
Douglas et al.: Lesquerella douglasii S. Wats. (Columbia Bladderpod)

Smith: 121. Pachystima myrsinites Raf. (September 6, 1926)
Douglas et al.: Pachistima myrsinites (Pursh) Raf. (Falsebox, Mountain or Oregon Boxwood, Mountain-box, or Mountain-lover)

Smith: 122. Fragaria glauca (Wats.) Rydb. (September 6, Nash)
Douglas et al.: Fragaria virginiana Duch. ssp. glauca (S. Wats.) Staudt (Wild Strawberry)

Smith: 123. Artemisia discolor Dougl. (Kitwanga, BC, first terrace August 28, 1926)
Douglas et al.: Artemisia michauxiana Bess. in Hook. (Michaux's Mugwort)

Smith: 124. Lycopodium complanatum L.
Douglas et al.: Lycopodium complanatum L. (Ground-cedar)

Smith: 125. Cladonia sp. (first terrace Kitwanga, August 28, 1926)

Smith: 126. Geranium Richardsonii F. and M.
Douglas et al.: Geranium richardsonii Fisch. & Trautv. (White or Richardson's Geranium or Crane's-bill)

Smith: 127. Pyrola asarifolia Michx.
Douglas et al.: Pyrola asarifolia Michx. var. asarifolia, P. asarifolia Michx. var. purpurea (Bunge) Fern. (Pink, Large, or Common Pink Wintergreen)

Smith: 128. Sonchus arvensis L. var. maritimus (L.) Wg.
Douglas et al.: Sonchus arvensis L. (Sow-thistle)

Smith: 129. Sonchus arvensis L. var. maritimus (L.) Wg.
Douglas et al.: Sonchus arvensis L. (Sow-thistle)

Smith: 130. Hosackia denticulata Drew.
Douglas et al.: Lotus denticulatus (Drew) Greene (Meadow Bird's-foot Trefoil)

Smith: 131. Habenaria obtusata Richards.
Douglas et al.: Platanthera obtusata (Banks ex Pursh) Lindl. (One-leaved Rein Orchid)

Smith: 132. Polypodium vulgare L. var. occidentale (Hook.)
Douglas et al.: Polypodium glycyrrhiza D.C. Eaton (Licorice Fern)

Smith: 133. Campanula rotundifolia L.
Douglas et al.: Campanula rotundifolia L. var. alaskana A. Gray, C. rotundifolia L. var. rotundifolia (Common Harebell, or Bluebells of Scotland)

Smith: 134. Senecio triangularis Hook. (Dwarf)?
Douglas et al.: Senecio triangularis Hook. (Arrow-leaved Groundsel or Ragwort)

Smith: 135. Agrimonia gryposepala Wallr.
Douglas et al.: Agrimonia gryposepala Wallr. (Common Agrimony)

Smith: 136. Angelica genuflexa Nutt.
Douglas et al.: Angelica genuflexa Nutt. (Kneeling Angelica)

Smith: 137. Spiraea Douglasii Hook. var. Menziesii Presl.
Douglas et al.: Spiraea douglasii Hook. ssp. menziesii (Hook.) Calder & Taylor (Hardhack, Pink Spiraea or Menzies' Spirea)

Smith: 138. Ribes oxyacanthoides L.

Douglas et al.: Ribes oxyacanthoides L. ssp. cognatum Greene, R. oxyacanthoides ssp. oxyacanthoides (Northern or Northern Smooth Gooseberry)

Smith: 139. Clematis columbiana Hornem.

Douglas et al.: Clematis occidentalis (Hornem.) DC. ssp. grosseserrata (Rydb.) Taylor & MacBryde (Columbia or Blue Clematis or Virgin's Bower)

Smith: 140. Astragalus alpinus L.

Douglas et al.: Astragalus alpinus L. (Alpine Milk-vetch)

Smith: 141. ["large fungus like sponge beautiful"] (from Hazelton, B.C., September 12, 1926)

Smith: 142. Neslia paniculata (L.) Desv. (September 5, 1926)

Douglas et al.: Neslia paniculata (L.) Desv. (Ball Mustard)

Smith: 143. Angelica genuflexa Nutt. (September 5, 1926)

Douglas et al.: Angelica genuflexa Nutt. (Kneeling Angelica)

Smith: 144. Ledum groenlandicum Oeder. (August 29, 1926, muskeg near lake near hospital [at] Hazelton)

Douglas et al.: Ledum groenlandicum Oeder (Labrador Tea)

Smith: 145. [no data provided by Smith, but this number seems to refer to Sphagnum sp.)

Smith: 146. Parnassia palustris L.

Douglas et al.: Parnassia palustris L. (Northern Grass-of-Parnassus)

Smith: 147. Aster junceus Ait. (September 5, 1926)

Douglas et al.: ?Aster borealis (T. & G.) Prov. (Rush or Boreal Aster)

Smith: 148. Mentha canadensis L.

Douglas et al.: Mentha arvensis L. (Field Mint)

Smith: 149. Dracocephalum parviflorum Nutt. (September 5, 1926)

Douglas et al.: Dracocephalum parviflorum Nutt. (American Dragonhead)

Smith: 150. Ranunculus repens L. (August 14, 26, slough south side of Skeena Kitwanga)

Douglas et al.: Ranunculus repens L. (Creeping or Swamp Buttercup)

Smith: 151. Delphinium Brownii Rydb.

Douglas et al.: ?Delphinium glaucum S. Wats. (Tall Larkspur)

Smith: 152. Corallorhiza Mertensiana Bong. (September 5, 1926)

Douglas et al.: Corallorhiza maculata Raf. ssp. mertensiana (Bong.) Calder & Taylor (Western Coralroot)

Smith: 153. Tiarella unifoliata Hook. (September 5, 1926)

Douglas et al.: Tiarella trifoliata L. (Cut-leaved Foamflower)

Smith: 154. Chimaphila umbellata (L.) Nutt. (September 5, 1926)

Douglas et al.: Chimaphila umbellata (L.) Bart. ssp. occidentalis (Rydb.) Hult. (Prince's Pine or Pipsissewa)

Smith: 155. Lycopodium obscurum L. var. dendroideum Eaton (September 5, 1926)

Douglas et al.: Lycopodium dendroideum Michx. (Ground-pine)

Smith: 156. Menziesia ferruginea Sm. (September 5, 1926)

Douglas et al.: Menziesia ferruginea Sm. ssp. ferruginea (False Azalea)

Smith: 157. Pachystima myrsinites Raf. (September 5, 1926)

Douglas et al.: Pachistima myrsinites (Pursh) Raf. (Falsebox, Mountain or Oregon Boxwood, Mountain-box, or Mountain-lover)

Smith: 158. Corydalis aurea Willd. (July 31, 1926, Kitwanga)

Douglas et al.: Corydalis aurea Willd. ssp. aurea (Golden Corydalis)

Smith: 159. Mentha canadensis L. (no number) (July 31, 1926, Kitwanga)

Douglas et al.: Mentha arvensis L. (Field Mint)

Smith: 160. Vaccinium caespitosum Michx. (July 23, 1926, Kitwanga)

Douglas et al.: Vaccinium caespitosum Michx. (Dwarf Blueberry or Bilberry)

Smith: 161. Chelone nemorosa Dougl.
Douglas et al.: Chelone nemorosa Dougl. (Turtlehead)

Smith: 162. Aruncus sylvester Kost. (June 25, 1926)
Douglas et al.: Aruncus dioicus (Walt.) Fern. (Goatsbeard)

Smith: 163. Solidago lepida DC.
Douglas et al.: Solidago canadensis L. (Canada Goldenrod)

Smith: 164. [shelf fungus, last of 1926 collections] (from Hazelton, B.C., August 15, 1926)

Additional specimens:
Smith: 182. [no data provided by Smith, but this number refers to Crataegus douglasii]
Douglas et al.: Crataegus douglasii Lindl. (Black Hawthorn)

Smith: 183. [no data provided by Smith, but this number refers to Amelanchier alnifolia]
Douglas et al.: Amelanchier alnifolia (Nutt.) Nutt. (saskatoon)

Smith: 242. [no data provided by Smith, but this number refers to Apocynum androsaemifolium]
Douglas et al.: Apocynum androsaemifolium L. (Spreading Dogbane)

Appendix 3. Taxa reported by H.I. Smith in individual species accounts for which no observations or collections were reported.

(1) Identified by Smith as: fungus on birch or hemlock [see main text, #2]

(2) Identified by Smith as: shelf fungus [see Appendix 4, #1]

(3) ?Athyrium filix-femina (Lady Fern), identified by Smith as: unidentified fern; "a fern, (collect)"

(4) Chamaecyparis nootkatensis (D. Don in Lamb.) Spach (Yellow Cedar or Cypress, or Alaska Cedar or Cypress), identified by Smith as: Chamaecyparis nootkatensis (Lamb) Spach. (Yellow Cypress)

(5) Thuja plicata Donn ex D. Don in Lamb. (Western Redcedar), identified by Smith as: Thuja plicata Donn. (Red Cedar)

(6) Identified by Smith as: Abies sp. (Balsam Fir) (which of 3?)

(7) Picea x lutzii Little (Hybrid Sitka Spruce), identified by Smith as: Picea sitchensis Carr (Sitka Spruce)

(8) Pinus contorta Dougl. ex Loud. var. latifolia Engelm. (Lodgepole Pine), identified by Smith as: Pinus contorta Dougl. (Scrub Pine)

(9) Pseudotsuga menziesii (Mirb.) Franco (Douglas fir), identified by Smith as: Pseudotsuga mucronata Raf. (Douglas Fir)

(10) Tsuga heterophylla (Raf.) Sarg. (Western Hemlock), identified by Smith as: Tsuga heterophylla Sarg. (Western Hemlock)

(11) Taxus brevifolia Nutt. (Western or Pacific Yew), identified by Smith as: Taxus brevifolia Nutt. (Yew)

(12) Lysichiton americanum Hult. & St. John (Skunk-cabbage), identified by Smith as: Lysichiton kamtschatcense Schott. (Skunk Cabbage)

(13) Identified by Smith as: Allium sp. (Wild Onion)

(14) Identified by Smith as: "grass"

(15) Identified by Smith as: (242) Apocynum sp. (Spreading Dogbane)

(16) Oplopanax horridus (Smith) Miq. (Devil's club), identified by Smith as: Fatsia horrida B. & H. (Devil's Club)

(17) Betula papyrifera Marsh. var. papyrifera (Paper, White or Canoe Birch), identified by Smith as: Betula sp. (Birch)—There is only one kind of birch in the Gitksan country ([Luke] Fowler)

(18) Identified by Smith as: Ribes bracteosum

(19) Nuphar polysepalum Engelm. (Rocky Mountain Cow-lily, Spatterdock, Yellow Pond-lily), identified by Smith as: ?Nymphaea polysepala (Engelm.) Greene. ("Water Lily")

(20) Populus tremuloides Michx. (Trembling Aspen), identified by Smith as: <u>Populus tremuloides</u> Michx. (Trembling Aspen)

(21) <u>Populus</u> <u>balsamifera</u> L. ssp. <u>trichocarpa</u> (T. & G.) Brayshaw (Black Cottonwood) and possibly <u>P</u>. <u>balsamifera</u> L. ssp. <u>balsamifera</u> (Balsam Poplar), identified by Smith as: <u>Populus</u> <u>trichocarpa</u> T. & B. (Black Cottonwood)

(22) †<u>Nicotiana</u> <u>tabacum</u> L. (Tobacco)

Appendix 4. Botanical species that lack Gitksan names or uses, but which H.I. Smith included in the main body of his original manuscript.

Fungi (Mushrooms and Their Relatives)

Unidentified Family (?Polyporaceae)

(1) Unidentified Fungus
Identified by Smith as: (164) "shelf fungus"

Editorial comments: Smith associated this "shelf fungus" with "Bob Robinson, September 5, 1926," but offered no further details regarding its botanical identity, or Gitksan name or use.

Lichens (Lichenized Fungi)

Cladoniaceae

(2) ?Cladonia bellidiflora (Ach.) Schaer. (Red Pixie Cups) and, perhaps, other Cladonia spp.
Identified by Smith as: (125) Cladonia sp. (red top lichen or moss?) (from first terrace, Kitwanga, B.C., August 28, 1926)[196]

Among the Gitksan, according to Gus Sampare, September 30, 1926, this plant was of no use and had no name.

Pteridophytes (Ferns and Their Relatives)

Lycopodiaceae (Clubmoss Family)

(3) ?Lycopodium dendroideum Michx. (Ground-pine)
Identified by Smith as: (115) Lycopodium obscurum L. (Club Moss, Ground Pine)

Among the Gitksan, according to Luke Fowler, May 24, 1926, this plant was of no use.

---

[196] This may correspond to Smith's collection number 61.

Polypodiaceae (Polypody or Common Fern Family)

(4) Polypodium glycyrrhiza D.C. Eaton (Licorice Fern)
Identified by Smith as: (132) Polypodium vulgare L. var occidentale (Hook) (Polypody)

According to Bob Robinson, October 3, 1926, it is too dry to be able to tell what it is.

Gymnosperms (Conifers and the Taxad, Western Yew)

Pinaceae (Pine Family)

(5) Pseudotsuga menziesii (Mirb.) Franco (Douglas Fir)[197]
Identified by Smith as: Pseudotsuga mucronata Raf. (Douglas Fir)[198]

Among the Gitksan, according to Luke Fowler, July 8, 1926, the Douglas fir was never used because not one of these trees is found in the Gitksan country.

Angiosperms (Flowering Plants), Dicotyledons

Asteraceae (syn. Compositae, Aster or Composite Family)

(6) Arnica cordifolia Hook. (Heart-leaved Arnica)
Identified by Smith as: (108) Arnica cordifolia Hook. (Dwarf Sunflower)

Among the Gitksan, according to Luke Fowler, June 8, 1926, this plant was of no use and had no name.

(7) ?Aster borealis (T. & G.) Prov. (Rush or Boreal Aster)
Identified by Smith as: (147) Aster junceus Ait. (Aster)

---

[197] The distribution of this tree within British Columbia does not extend as far north as Gitksan territory (Krajina et al. 1982).
[198] Elsewhere, Smith (1929:51) identified this species with the synonym Pseudotsuga taxifolia Britt.

Among the Gitksan, according to Bob Robinson, September 5, 1926, this plant was of no use and had no name.

(8) <u>Aster</u> sp. (Aster)
Identified by Smith as: (26) <u>Aster</u> sp. (Blue Aster)

Among the Gitksan, according to both Abraham and John Fowler, August 9th and September 16, 1925, this plant was of no use. John said it was good for cows[199] to eat but not for deer.[200]

(9) <u>Cirsium undulatum</u> (Nutt.) Spreng. (Wavy-leaved Thistle)
Identified by Smith as: (117) <u>Cirsium</u> (<u>Carduus</u>) <u>undulatum</u> (Nutt.) Spring. (Wooly Thistle)

Among the Gitksan, according to Bob Robinson, September 5, 1926, this plant was of no use and had no name. He said it was introduced by white men.

(10) <u>Crepis elegans</u> Hook. (Elegant Hawksbeard)
Identified by Smith as: (52D) <u>Crepis elegans</u> Hook. (Hawk's Beard) (from near Woodcock, B.C., May 29, 1926)

Among the Gitksan, according to Abraham Fowler, May 29, 1926, this plant was of no use and had no name.

(11) <u>Erigeron philadelphicus</u> L. (Philadelphia Fleabane or Daisy)
Identified by Smith as: (52G) <u>Erigeron philadelphicus</u> L. (Common Fleabane) (from Hazelton, B.C., May 24, 1926; from near Woodcock, B.C., May 29, 1926)

Among the Gitksan, according to Luke Fowler, May 24, 1926 and Abraham Fowler, May 29, 1926, this plant was of no use and had no name.

(12) <u>Senecio cymbalarioides</u> Buek non Nutt. (Alpine Meadow Butterweed)
Identified by Smith as: (72) <u>Senecio cymbalarioides</u> Nutt. (Yellow Composite)

---

[199] The domesticated cow (<u>Bos taurus</u> Linnaeus) is known by the Gitksan by the Chinook Jargon term, mismuus.
[200] Mule deer (<u>Odocoileus hemionus</u> [Rafinesque]) is known in Gitksan as wan.

Among the Gitksan, according to Luke Fowler, June 15, 1926, this plant was of no use and had no name.

(13) Senecio pauperculus Michx. (Canadian Butterweed)
Identified by Smith as: (52E) Senecio balsamitae Muhl. (from near Woodcock, B.C., May 29, 1926)

Among the Gitksan, according to Abraham Fowler, May 29, 1926, this plant was of no use and had no name.

(14) Senecio triangularis Hook. (Arrow-leaved Groundsel or Ragwort)
Identified by Smith as: (134) Senecio triangularis Hook. (Groundsel, Ragwort) (Dwarf)

Bob Robinson, October 3, 1926, said he could not tell this plant as it was too dry.

(15) Solidago canadensis L. (Canada Goldenrod)
Identified by Smith as: (28) (163) Solidago lepida DC. (Goldenrod)

Among the Gitksan, according to Abraham Fowler, August 9, 1925, this plant was of no use. According to Bob Robinson, 1926 this plant was of no use and had no name.

(16) †Taraxacum officinale Weber in Wiggers (Common Dandelion)[201] and possibly also T. ceratophorum (Ledebour) A.P. de Candolle (Horned Dandelion)
Identified by Smith as: Taraxacum sp. (Dandelion)
Gitksan name: Yench
Modern spelling of Gitksan name: 'yens (WG), 'yans (EG) (literally, 'leaf')[202]

Among the Gitksan, according to Luke Fowler, May 24, 1926, this plant was of no use. It was just considered pretty.

Boraginaceae (Borage Family)

(17) Mertensia paniculata (W. Ait.) G. Don (Tall Bluebells, or Panicled Mertensia)

---

[201]This is an introduced plant.
[202]This term refers to plant leaves as well as to economically useless plant such as most grasses or weeds (Hindle and Rigsby 1973).

Identified by Smith as: (54) <u>Mertensia paniculata</u> Don. (Lungwort; Bluebell plant like Fox Glove)[203] (from Hazelton, B.C., June 6, 1926)

Among the Gitksan, according to Luke Fowler, June 8, 1926, this plant was of no use and had no name.

Brassicaceae (syn. Cruciferae, Mustard or Crucifer Family)

(18) <u>Arabis hirsuta</u> (L.) Scop. (Hairy Rockcress)
Identified by Smith as: <u>Arabis hirsuta</u> (L.) Scop. (a Rock Cress, Tall Mustard) (from near Woodcock, B.C. May 29, 1926) (Smith collection no. 107)

Among the Gitksan, according to Abraham Fowler, May 29, 1926, and Luke Fowler, June 8, 1926, this plant was of no use and had no name.

(19) <u>Capsella bursa-pastoris</u> (L.) Medic. (Shepherd's purse)
Identified by Smith as: (105) <u>Capsella bursa-pastoris</u> (L.) Medic. (Shepherd's Purse)

Among the Gitksan, according to Luke Fowler, June 8, 1926, this plant was of no use and had no name.

(20) ?<u>Draba aurea</u> Vahl in Horn. (Golden Draba or Whitlow-grass)
Identified by Smith as: (113) <u>Draba lutea</u> Gilib (looks like Shepherd's Purse but is yellow)

Among the Gitksan, according to Luke Fowler, May 24, 1926, this plant was of no use and had no name in their language.

(21) <u>Erysimum cheiranthoides</u> L. (Wormseed Mustard)
Identified by Smith as: (74) <u>Erysimum cheiranthoides</u> L. (Wormseed Mustard)

Among the Gitksan, according to Luke Fowler, June 15, 1926, and according to other information obtained in 1926, this plant was of no use and had no name.

(22) <u>Lesquerella douglasii</u> S. Wats. (Columbia Bladderpod)

---

[203] Fox glove is the common name for <u>Digitalis</u> spp.

Identified by Smith as: (120) <u>Lesquerella</u> <u>Douglasii</u> Wats. (Bladder-pod, Sage) (from Nash, September 6, 1926)

Bob Robinson said to collect this again as he could not tell what it was from a dry specimen.

(23) <u>Neslia</u> <u>paniculata</u> (L.) Desv. (Ball Mustard)
Identified by Smith as: (142) <u>Neslia</u> <u>paniculata</u> (L.) Desv. (Ball Mustard)

Among the Gitksan, according to Bob Robinson, September 5, 1926, this plant was of no use, had no name and no story about it.

Ericaceae (Heath Family)

(24) <u>Chimaphila</u> <u>umbellata</u> (L.) Bart. ssp. <u>occidentalis</u> (Rydb.) Hult. (Prince's Pine or Pipsissewa)
Identified by Smith as: (154) <u>Chimaphila</u> <u>umbellata</u> (L.) Nutt. (Prince's Pine)

Among the Gitksan, according to Bob Robinson, September 5, 1926, this plant was of no use and had no name.

(25) <u>Monotropa</u> <u>uniflora</u> L. (Indian-pipe)
Identified by Smith as: (6) <u>Monotropa</u> <u>uniflora</u> L. (Indian Pipe) (from terrace near falls, south side of Skeena River, about two miles east of Kitwanga, B.C., August 23, 1925)

Among the Gitksan, according to John Fowler, August 27, 1925, this plant was of no use.

(26) <u>Pyrola</u> <u>asarifolia</u> Michx. var. <u>asarifolia</u> and/or <u>P</u>. <u>asarifolia</u> Michx. var. <u>purpurea</u> (Bunge) Fern. (Pink, Large, or Common Pink Wintergreen)
Identified by Smith as: (127) <u>Pyrola</u> <u>asarifolia</u> Michx. (Wintergreen) (from Kitwanga, B.C.)

Among the Gitksan, according to Bob Robinson, October 3, 1926, this common plant was of no use—not even the roots—and had no name.

Fabaceae (syn. Leguminosae, Bean or Legume Family)

(27) <u>Astragalus</u> <u>alpinus</u> L. (Alpine Milk-vetch)
Identified by Smith as: (140) <u>Astragalus</u> <u>alpinus</u> L. (Milk Vetch)

Among the Gitksan, according to Bob Robinson, October 3, 1926, this plant was of no use and had no name.

(28) <u>Lotus</u> <u>denticulatus</u> (Drew) Greene (Meadow Bird's-foot Trefoil)
Identified by Smith as: (130) <u>Hosackia</u> <u>denticulata</u> Drew. (Bird-foot Clover) (from railway
    between Nash and Andimaul, B.C., September 1, 1926)

Among the Gitksan, according to Bob Robinson, September 5, 1926, and October 3, 1926, this plant was of no use and had no name.

(29) <u>Oxytropis</u> <u>monticola</u> A. Gray ssp. <u>monticola</u> (Mountain Locoweed)
Identified by Smith as: (57) <u>Oxytropis</u> <u>monticola</u> Gray (Loco-Weed)[204]

Among the Gitksan, according to Luke Fowler this plant was of no use and had no name.

Fumariaceae (Fumitory Family)

(30) <u>Corydalis</u> <u>aurea</u> Willd. ssp. <u>aurea</u> (Golden Corydalis)
Identified by Smith as: (109) (158) <u>Corydalis</u> <u>aurea</u> Willd. (Golden Corydalis, "Yellow
    Dutchman's Breeches") (from Kitwanga, B.C., July 31, 1926)
Gitksan name: (just) mezeroolE = flower
Modern spelling of Gitksan name: majagalee (literally, 'flower')

Among the Gitksan, according to Luke Fowler, June 8, 1926, and two women, July 31, 1926, this plant was of no use and had no name other than mezerulē, flowers.[205]

Geraniaceae (Geranium Family)

---

[204] Elsewhere Smith associated collection number 57 with <u>Actaea</u> <u>eburnea</u> (see Appendix 2, #57).

[205] The Gitksan consultants who worked with Smith gave the general term meaning 'flower' for this species as well as for <u>Parnassia</u> <u>palustris</u> (Appendix 4, #34, #35), <u>Clematis</u> <u>occidentalis</u> (Appendix 4, #41) and <u>Heuchera</u> <u>glabra</u> (Appendix 4, #45). This 'flower' term does not refer to any species in particular, but rather to any flower or small, essentially useless plant with conspicuous flowers.

(31) <u>Geranium</u> <u>erianthum</u> DC. (Northern Geranium or Crane's-bill)
Identified by Smith as: (104) <u>Geranium</u> <u>erianthum</u> DC. (Northern Geranium)

Among the Gitksan, according to Luke Fowler, June 8, 1926, this plant was of no use and had no name.

Lamiaceae (syn. Labiatae, Mint Family)

(32) <u>Dracocephalum</u> <u>parviflorum</u> Nutt. (American Dragonhead)
Identified by Smith as: (149) <u>Dracocephalum</u> <u>parviflorum</u> Nutt. (Dragon Head)

Among the Gitksan, according to Bob Robinson, September 5, 1926, this plant was of no use and had no name. He says to collect a fresh specimen to verify this.

Menyanthaceae (Buckbean Family)

(33) <u>Menyanthes</u> <u>trifoliata</u> L. (Buckbean)
Identified by Smith as: (56) <u>Menyanthes</u> <u>trifoliata</u> L. (Buckbean) (from pond west of road
    between Hazelton station and Hazelton, B.C. June 8, 1926)

Among the Gitksan, according to Luke Fowler, this plant was of no use and had no name.

Parnassiaceae (Grass-of-Parnassus Family)

(34) <u>Parnassia</u> <u>palustris</u> L. (Northern Grass-of-Parnassus)
Identified by Smith as: (83) <u>Parnassia</u> <u>montanensis</u> Rydb. & Fer. (Mountain Grass of
    Parnassus)
Gitksan name: Skan mizerlay skan meaning plant, mizerlay flowers
Modern spelling of Gitksan name: sganmajagalee (literally, "flower-plant")[206]

Among the Gitksan, according to Bob Robinson, July 24, 1926, this plant was of no use.

(35) <u>Parnassia</u> <u>palustris</u> L. (Northern Grass-of-Parnassus)

---

[206]This is probably a nonce form.

Identified by Smith as: (146) <u>Parnassia palustris</u> L. (Marsh Grass of Parnassus) (from muskeg near Hazelton, B.C.)

Among the Gitksan, according to Bob Robinson, September 5, 1926, this plant was of no use and had no name.

Polemoniaceae (Phlox Family)

(36) ?<u>Collomia linearis</u> Nutt. (Narrow-leaved Collomia)
Identified by Smith as: (96) <u>Gilia linearis</u> Gray (from Kitwanga, B.C., June 21, 1926)

Among the Gitksan, according to Abraham Fowler, June 21, 1926, this plant was of no use.

(37) <u>Microsteris gracilis</u> (Hook.) Greene (Pink Microsteris)
Identified by Smith as: (60) <u>Gilia gracilis</u> Hook. (from Hazelton, B.C., June 8, 1926)

Among the Gitksan, according to Luke Fowler, June 8, 1926, this plant was of no use and had no name.

(38) <u>Polemonium pulcherrimum</u>Hook. ssp. <u>pulcherrimum</u> (Showy Jacob's-ladder)
Identified by Smith as: (52H) <u>Polemonium pulcherrimum</u> Hook. (Greek Valerian) (from near Woodcock, B.C., May 29, 1926) (not in Henry's flora)

Among the Gitksan, according to Abraham Fowler, May 29, 1926, this plant was of no use and had no name.

Polygonaceae (Buckwheat Family)

(39) <u>Polygonum convolvulus</u> (L.) (Black Bindweed)
Identified by Smith as: (97) <u>Polygonum convolvulus</u> L. (Bindweed) (from Kitwanga, B.C., June 21, 1926)

Among the Gitksan, according to Abraham Fowler, June 21, 1926, this plant was of no use.

Primulaceae (Primrose Family)

(40) Trientalis arctica Fisch. ex Hook. (Northern Starflower)
Identified by Smith as: (61) Trientalis arctica Fisch. (Chickweed, Wintergreen) (from Hazelton, B.C., June 8, 1926)

Among the Gitksan, according to Luke Fowler, June 8, 1926, this plant was of no use and had no name.

Ranunculaceae (Buttercup Family)

(41) Clematis occidentalis (Hornem.) DC. ssp. grosseserrata (Rydb.) Taylor & MacBryde (Columbia or Blue Clematis or Virgin's Bower)
Identified by Smith as: (139) Clematis columbiana Hornem.
Gitksan name: MezerlE, flowers
Modern spelling of Gitksan name: majagalee (literally, 'flower')[207]

Among the Gitksan, according to Bob Robinson, October 3, 1926, this plant was a native, not introduced, was of no use, and had no name other than that applied to any flower or fancy thing.

(42) Ranunculus repens L. (Creeping or Swamp Buttercup)
Identified by Smith as: (150) Ranunculus repens L. (Creeping Buttercup) (from bank of slough south of Skeena River, near Kitwanga, B.C.)

Among the Gitksan, according to Mr. and Mrs. Robert A. Sampare, August 14, 1926, and Bob Robinson, September 5, 1926, this plant was of no use and had no name.

Rosaceae (Rose Family)

(43) Potentilla gracilis Dougl. (Graceful Cinquefoil)
Identified by Smith as: (70) Potentilla viridescens Rydb. (Buttercup) (not in Henry's flora)

---

[207] This is probably a nonce form.

Among the Gitksan, according to Luke Fowler, June 15, 1926, this plant was of no use and had no name.

Rubiaceae (Madder Family)

(44) Galium boreale L. (Northern Bedstraw)
Identified by Smith as: (71) (95) Galium boreale L. (Bedstraw) (from Kitwanga, B.C., June 21, 1926)

Among the Gitksan according to Luke Fowler, June 15, 1926, and Abraham Fowler, June 21, 1926, this plant was of no use and had no name.

Saxifragaceae (Saxifrage Family)

(45) Heuchera glabra Willd. ex Roem. & Schult. (Smooth Alumroot)
Identified by Smith as: (90) Heuchera glabra Willd. (Alum Root) (Bishops Cap?)
Gitksan name: skan mezerulē, skan meaning plant and mezerulē blossom
Modern spelling of Gitksan name: sganmajagalee (literally, "flower-plant")[208]

Among the Gitksan, according to Bob Robinson, July 24, 1926, this plant was of no use.

(46) Saxifraga tricuspidata Rottb. (Three-toothed Saxifrage)
Identified by Smith as: (52F) Saxifraga tricuspidata Rottb. (from near Woodcock, B.C., May 29, 1926)

Abraham Fowler, May 29, 1926, this plant was of no use and had no name.

(47) Tiarella trifoliata L. (Cut-leaved Foamflower)
Identified by Smith as: (153) Tiarella unifoliata Hook. (Simply Leaved Tiarella, False Mitrewort)

Among the Gitksan, according to Bob Robinson, September 5, 1926, this plant was of no use and had no name.

---

[208] This is probably a nonce form.

Scrophulariaceae (Figwort Family)

(48) Collinsia parviflora Doug. ex Lindl. (Small-flowered Blue-eyed Mary)
Identified by Smith as: (59) Collinsia parviflora Dougl. (Blue-eyed Mary) (from Hazelton, B.C., June 8, 1926)

Among the Gitksan, according to Luke Fowler, June 8, 1926, this plant was of no use and had no name.

(49) ?Chelone nemorosa Dougl. (Turtlehead)[209]
Identified by Smith as: (161) Chelone nemorosa Dougl. (Turtlehead)

Among the Gitksan, according to Abraham Fowler, June 25, 1926, this plant was of no use and had no name, but older people might know.

(50) Rhinanthus minor L. (Yellow Rattle)
Identified by Smith as: (98) Rhinanthus crista-galli L. (Yellow Rattle) (from Kitwanga, B.C., June 21, 1926)

Among the Gitksan, according to Abraham Fowler, June 21, 1926, this plant was of no use.

Violaceae (Violet Family)

(51) Viola adunca J.E. Smith in Rees (Early Blue Violet)
Identified by Smith as: (47) Viola adunca Sm. (Violet) (from near Hazelton, B.C., 1926)

Among the Gitksan, according to Luke Fowler, May 24, 1926, this plant was of no use and had no name.

(52) Viola canadensis L. ssp. rydbergii (Greene) House in Rydb. (Canada Violet)
Identified by Smith as: (49) Viola canadensis L. (Violet) (from near Hazelton, B.C., 1926)

---

[209]This plant does not appear in Henry (1915) or in contemporary botanical works dealing with British Columbian flora (e.g., Douglas et al. 1991, Hitchcock and Cronquist 1973, Taylor and MacBryde 1977) and may represent an erroneous botanical identification.

Among the Gitksan, according to Luke Fowler, May 24, 1926, this plant was of no use and had no name.

Orchidaceae (Orchid Family)

(53) Corallorhiza maculata Raf. ssp. mertensiana (Bong.) Calder & Taylor (Western Coralroot) Identified by Smith as: (152) Corallorhiza Mertensiana Bong. (Coral Root)

Among the Gitksan, according to Bob Robinson, September 5, 1926, this plant was of no use and had no name. He said to collect a fresh specimen to verify this as this specimen is too dry.

Index

Abies amabilis 36, 150, 153, 155, 164
Abies lasiocarpa 36, 150, 153, 155
Abies sp. 36, 188
Acer glabrum 47, 153, 159, 175
Aceraceae 47
Achillea lanulosa 57, 174
Achillea millefolium 57, 155, 174
Actaea arguta 107, 179, 182
Actaea eburnea 177
Actaea rubra 107, 152, 159, 163, 177, 179, 182
Agrimonia gryposepala 109, 181, 184
Agrimonia striata 109, 159
agrimony 50
agrimony, common 181, 184
agrimony, grooved 109, 159
Agropyron sp. 148
alder 10
alder, green or Sitka 63, 153, 155
alder, mountain 60, 63, 150, 153, 155, 171
alder, red 34, 47, 60, 61, 63, 71, 150, 153, 164, 173
Alectoria 21
Alectoria jubata 21
Allium cernuum 140, 151
Allium sp. 140, 188
Alnus crispa 63, 153, 155
Alnus rubra 60, 150, 153, 164, 173
Alnus tenuifolia 60, 63, 150, 153, 155, 171
alumroot, smooth 180, 200
am'mal (EG) 129
am'mel (WG) 129
Amelanchier alnifolia 110, 150, 153, 172, 187
Amelanchier florida 110, 172
amhaawak 64
amhat'a'l (EG) 34
amhat'e'l (WG) 34
amk'ooxst 132
amluux 60
Anaphalis margaritacea 59, 157, 162, 179
Anemone globosa 107, 176, 182
Anemone multifida 107, 156, 176, 182
anemone, cut-leaved 107, 156, 176, 182
angelica 49
Angelica genuflexa 17, 49, 50, 109, 150, 152, 155, 157, 159, 162, 163, 182, 184, 185
angelica, kneeling 49, 50, 51, 75, 150, 152, 155, 157, 159, 162, 163, 182, 184, 185
angiosperms 47, 138, 149, 162, 191
Antennaria Howellii 182
Antennaria microphylla 176
Antennaria neglecta 182
Antennaria rosea 176
Apiaceae 49
Apocynaceae 53
Apocynum androsaemifolium 53, 153, 159, 160, 163, 187
Apocynum sp. 53, 188
apple, Pacific crab 115, 150, 153, 156, 173

Aquilegia formosa 108, 134, 150, 175
Arabis hirsuta 176, 182, 194
Araceae 138
Aralia nudicaulis 55, 152, 159, 163, 176
Araliaceae 55
Arctostaphylos uva-ursi 87, 150, 159, 163, 171, 172
Arnica cordifolia 182, 191
arnica, heart-leaved 182, 191
Artemisia discolor 59, 183
Artemisia michauxiana 59, 161, 183
arum family 138
Aruncus dioicus 112, 160, 179, 187
Aruncus sylvester 112, 179, 187
aspen 129
aspen, trembling 10, 132, 151, 154, 156, 164, 189
Asplenium cyclosorum 27, 175
Aster borealis 185, 191
Aster conspicuus 59, 161, 163, 179
aster family 57, 191
Aster junceus 185
Aster sp. 174, 192
aster, blue 174
aster, boreal 185, 191
aster, showy 59, 161, 163, 179
Asteraceae 57, 191
Astragalus alpinus 185, 196
Athyrium filix-femina 25, 27, 155, 175, 188
ax 27
azalea, false 91, 161, 186
Babine 79
baneberry 107, 152, 159, 163, 177, 179, 182
Barbeau, C.M. 103, 133
bean family 98, 195
bear 25, 34, 39, 43, 50, 55, 91, 96, 99, 103, 107, 113, 121, 122, 124, 130, 138, 141, 144, 163
bear, black 55, 71, 79, 107, 110, 112, 152, 159, 163
bear, grizzly 55, 71, 79, 110, 112, 152
bears 152
beaver 132, 160
bedstraw, northern 179, 181, 200
bedstraw, sweet-scented) 180
Bella Coola (also see Nuxalk) iii, 4, 13
berries, black 75
berry, black 98, 121, 144
Betula papyrifera 19, 63, 153, 160, 164, 188
Betula sp. 64, 188
Betulaceae 60
bilana 'watsx (EG) 29, 31, 109
bilberry, black 98, 102, 122
bilena 'watsx (WG) 29, 31, 109
bindweed, black 181, 198
birch 64, 188
birch family 60
birch, paper 10, 63, 153, 160, 164, 188
Blackfoot 11
bladderpod, Columbia 183, 194
blue-eyed Mary, small-flowered 178, 201
bluebells, tall 177, 193
blueberries 10
blueberry, Alaskan 93, 150

blueberry, black 98
blueberry, dwarf 75, 79, 96, 117, 150, 176, 179, 186
blueberry, high bush 93
blueberry, oval-leaved 93, 150
Board on Preservation of Totem Poles 11
borage family 193
Boraginaceae 193
Brassicaceae 68, 194
Brown, a Gitksan Indian at Hazelton, B.C. 14
bryophytes 22
buckbean 177, 197
buckbean family 197
buckwheat family 106, 198
bunchberry 10, 79, 96, 152, 172
buttercup family 107, 199
buttercup, creeping 186, 199
buttercup, kidney-leaved 108, 152, 159, 163, 176
butterweed, alpine meadow 179, 192
butterweed, Canadian 177, 193
Calla palustris 138, 156, 176
calla, wild 138, 156, 176
Campanula rotundifolia 68, 161, 163, 184
Campanulaceae 68
Canadian National Railway 4, 5, 11
Caprifoliaceae 68
Capsella bursa-pastoris 182, 194
Carrier iii, 4, 5, 11
carrot, Indian 103
Castilleja miniata 133, 134, 156, 181
Castor canadensis 160
cedar 61, 129, 133
cedar, red 34, 188
cedar, yellow 10, 31, 153, 188
Celastraceae 77
Cetraria 21
Cetraria juniperina 20, 21
Cetraria pinastri 20, 159
Cetraria spp. 20
cetraria, moonshine 20, 159
Chamaecyparis nootkatensis 31, 153, 188
Chelone nemorosa 187, 201
cherry, bitter 10, 117, 150, 172, 181
cherry, choke 49, 51, 96, 117, 119, 150, 153, 160, 171, 181
cherry, pin 150
cherry, pin or bird 117
cherry, wild 96
Chimaphila umbellata 186, 195
Chinese 99
chipmunk 79
chipmunk, yellow-pine 152
cinquefoil, graceful 179, 199
Cirsium (Carduus) undulatum 183
Cirsium undulatum 183, 192
Cladonia bellidiflora 190
Cladonia sp. 183
Cladonia spp. 190
Cladoniaceae 190
Clark, Frank 14
Clavaria 19
Clavariaceae 18

Clematis columbiana 185
Clematis occidentalis 185, 199
clematis, Columbia 185, 199
Clintonia uniflora 140, 158, 162, 181
club, devil's 10
clubmoss family 29, 190
clubmosses 10, 163
Collinsia parviflora 177, 178, 201
Collomia linearis 181, 198
collomia, narrow-leaved 181, 198
columbine, red 108, 150, 175
Comandra livida 133, 171
common fern family 191
composite family 57, 191
conifers 31, 191
conk, cinder 19, 155, 159
Corallorhiza maculata 186, 202
Corallorhiza Mertensiana 186
coralroot, western 186, 202
Cornaceae 79
Cornus canadensis 79, 152, 172
Cornus stolonifera 79, 150, 152, 153, 159, 172
Corydalis aurea 182, 186, 196
corydalis, golden 182, 186, 196
Corylus cornuta 66, 150, 153, 159, 175
Corylus rostrata 66, 175
cottonwood 22, 129, 132, 133
cottonwood, black 10, 129, 151, 154, 164, 189
cow 192
cow-parsnip 10, 49, 50, 51, 119, 150, 155, 159, 173
crab apple, Pacific 10
cranberry 121
cranberry, bog 10, 91, 150, 173
cranberry, highbush 10, 50, 75, 96, 150, 153, 155, 171
Crassulaceae 82
Crataegus brevispina 112, 173
Crataegus douglasii 112, 150, 153, 173, 187
Crepis elegans 176, 192
crow 163
crucifer family 68, 194
Cupressaceae 31
currant family 101
currant, northern black- 102, 150, 175, 180
currant, stink 101
currant, trailing black 101
currants 10
cypress family 31
cypress, yellow 31, 188
damtx (EG) 25, 36, 39, 146
dandelion, common 193
dandelion, horned 193
Davies, D.F. 15
deer 192
Delphinium Brownii 108, 179, 186
Delphinium glaucum 49, 50, 108, 109, 159, 163, 179, 186
demtx (WG) 25, 36, 39, 146
Dennstaedtiaceae 25
Department of Indian Affairs 14
devil's club 25, 55, 75, 146, 155, 188
dicotyledons 47, 149, 191

204

didihuxw 19
dilusa 102
dilusa'a 102
Disporum hookeri 141, 161, 180
Disporum oreganum 141, 180
Disporum trachycarpum 141, 161
dogbane 112
dogbane family 53
dogbane, spreading 53, 153, 159, 160, 163, 187, 188
dogwood family 79
dogwood, red-osier 10, 79, 150, 152, 153, 159, 172
Draba aurea 182, 194
Draba lutea 182
draba, golden 182, 194
Dracocephalum parviflorum 185, 197
dragonhead, American 185, 197
Dryas drummondii 113, 160, 183
Dryopteridaceae 25, 27
Dryopteris austriaca complex 27, 150
Dryopteris carthusiana 27
Dryopteris expansa 27
Elaeagnaceae 84
elder, coastal red 61, 71, 93, 96, 121, 150, 155, 174
elderberry, red 10
Elymus glaucus 148, 154, 178
Elymus sp. 178
Elymus trachycaulus 148, 154, 178
Epilobium angustifolium 106, 150, 153, 158, 159, 162, 173
Equisetaceae 27
Equisetum arvense 27, 153, 174
Equisetum hyemale 29, 161, 174, 175
Ericaceae 87, 162, 195
Erigeron philadelphicus 177, 182, 192
Erysimum cheiranthoides 179, 194
eulachon 39, 43, 71, 75, 87, 91, 96, 99, 103, 107, 110, 112, 113, 121, 122, 124, 141, 144
evening-primrose family 106
everlasting, pearly 162
Fabaceae 98, 195
fairybells, Hooker's 141, 161, 180
fairybells, rough-fruited 141, 161
falsebox 10, 77, 159, 163, 183, 186
Fatsia horrida 55, 188
fern 188
fern, bracken 25, 155
fern, lady 10, 25, 27, 155, 175, 188
fern, licorice 10, 184, 191
fern, spiny 27
fern, spreading wood 27
fern, toothed wood 27
fern, wood 10
ferns 25, 190
ferns, wood 150
Field Museum of Natural History 15
figwort family 133, 201
fir, amabilis 10, 36, 150, 153, 155, 164
fir, balsam 25, 34, 36, 146, 188
fir, Douglas 188, 191

fir, Pacific silver 36, 150, 153, 155
fir, subalpine 10, 36, 150, 153, 155
fireweed 10, 106, 109, 150, 153, 158, 159, 162, 173
fleabane, Philadelphia 177, 182, 192
flowering plants 47, 138
foamflower, cut-leaved 186, 200
food (animal) 144
food, secondary 36, 39, 41, 129, 132
food-related use 49, 50, 66, 79, 87, 124, 129
Fowler, Abraham 13, 57, 60, 63, 75, 79, 82, 108, 109, 112, 129, 134, 136, 140, 141, 192, 193, 194, 198, 200, 201
Fowler, John 13, 21, 49, 63, 73, 75, 79, 84, 85, 87, 91, 106, 110, 112, 115, 117, 119, 121, 124, 127, 133, 136, 144, 192, 195
Fowler, Luke 4, 13, 14, 20, 22, 25, 27, 29, 31, 32, 34, 36, 39, 41, 43, 47, 50, 53, 55, 57, 61, 64, 66, 68, 71, 75, 79, 87, 91, 98, 99, 101, 102, 103, 104, 106, 107, 108, 109, 110, 112, 113, 121, 122, 124, 125, 127, 129, 132, 133, 134, 136, 138, 140, 142, 144, 145, 146, 148, 188, 190, 191, 192, 193, 194, 196, 197, 198, 199, 200, 201, 202
fox 53
Fragaria glauca 113, 174, 183
Fragaria virginiana 113, 150, 165, 174, 183
Fritillaria camschatcensis 141, 151, 180
Fritillaria kamtschatcense 141
frog 49, 108, 109, 113, 124, 163
frogs and/or toads 49, 109, 124, 152, 159, 160
Fumariaceae 196
fumitory family 196
fungi 18, 149, 190
fungi, lichenized 190
fungus 19, 20, 161, 185, 187, 188, 190
gahldaats 104
Galium boreale 179, 181, 200
Galium triflorum 180
gam (EG) 110
ganlaxsga'nist 60
gapk'oyp 79
gasx 141
gayda ts'uuts' 20
gayda ts'uuts' (EG) 19
gayda ts'uuts' (WG) 19
gem (WG) 110
Geocaulon lividum 133, 158, 162, 165, 171
Geraniaceae 101, 196
Geranium erianthum 182, 197
geranium family 101, 196
Geranium richardsonii 101, 161, 183
geranium, northern 182, 197
geranium, white 101, 161, 183
giikw 43
giist 63
Gilia gracilis 178
Gilia linearis 181
ginseng family 55
goat 106
goatsbeard 112, 160, 179, 187
goldenrod, Canada 187, 193
gooseberries 10

gooseberry 103, 121
gooseberry family 101
gooseberry, black 102, 156
gooseberry, black or black swamp 101
gooseberry, northern 102, 150, 185
grass 99, 148, 151, 188
grass family 148
grass-of-Parnassus family 197
grass-of-Parnassus, northern 180, 185, 197
Grossulariaceae 101
ground-cedar 29, 159, 183
ground-pine 31, 159, 182, 186, 190
groundhog 39, 43, 53, 71, 91, 96, 99, 103, 107, 110, 112, 122, 141
groundsel, arrow-leaved 184, 193
grouse 73
gwiilahl ganaa'w 22
gymnosperms 31, 149, 162, 191
ha'mook 50
ha'mookhl ganaa'w 49, 108, 109
haast 106
habasxw 148
Habenaria obtusata 148, 184
hagimgasxw 98
Haisla iii
haluuts'ook' 119
hardhack 129, 161, 181, 183, 184
harebell family 68
harebell, common 68, 161, 163, 184
hat'a'l (EG) 34
hat'e'l (WG) 34
hawksbeard, elegant 176, 192
hawthorn, black 10, 61, 112, 150, 153, 173, 187
haxwdakw 45
hay-scented fern family 25
hazel 66
hazelnut, beaked 10, 66, 150, 153, 159, 175
heath family 87, 195
hellebore, Indian 156, 162
hellebore, Indian or green false 145
hemlock 34, 41, 43
hemlock, western 10, 43, 142, 153, 164, 188
Henaaksiala iii
Heracleum lanatum 49, 50, 109, 150, 155, 159, 173
Hericium coralloides 19
Heuchera glabra 180, 200
hisgantxwit 59, 129
hishaasxwit 109
hishaawakxwit 113
hishabasxwit 68
hishinakxwit 148
hisk'awtsxwit 103
hismaawint 29
hisnisk'o'otxwit 101
hissgant'imi'ytsit, hissgant'imi'ytxwit 77
hla'anisihl sginist 20
hlingit 127
honeysuckle family 68
hoo'oxs 36
hoobixs 'Wii Gat (EG) 140
hoobixs 'Wii Get (WG) 140
horse 108
horsetail 121

horsetail family 27
horsetail, common 27, 47, 153, 174
horsetail, field 27, 153, 174
horsetails 10
Hosackia denticulata 184
huckleberries, black 71, 93
huckleberries, blue 71, 93
huckleberry 144
huckleberry, black 98, 150, 165
huckleberry, blue 93
Hydnaceae 18
Hydnum 19
ihlee'em ts'ak 108, 134
Indian hellebore 25, 39, 146
Indian-pipe 171, 195
Inonotus obliquus 19, 155, 159, 162
is 84
isimganaa'wtxws 49
Jacob's-ladder, showy 177, 198
jay, blue 79
juniper 32
juniper, common 31, 153, 155, 178
juniper, ground 31, 153, 155, 178
Juniperus communis 31, 153, 155, 178
Juniperus sp. 31, 178
k'alamst 121
k'ooxst 47
Kemano iii
kinnikinnick 10, 87, 150, 159, 163, 171, 172
Kitamaat iii
Kitlope iii
laam 107
Lactuca biennis 59, 161, 163, 180
Lactuca spicata 59, 180
Laknitz, Jim 14
Laknitz, Johnson 13, 22, 87
Lamiaceae 103, 197
larkspur, tall 108, 159, 163, 179, 186
Lathyrus ochroleucus 98, 150, 175
Ledum groenlandicum 89, 150, 155, 162, 185
legume family 98, 195
Lesquerella Douglasii 183, 194
Letharia vulpina 21, 41
lettuce, tall blue 59, 161, 163, 180
lichen 171, 172, 174, 190
lichen, lung 159
lichen, wolf 41
lichenized fungi 20
lichens 149, 190
lichens, lung 22
Liliaceae 140, 162
lily family 140
lily, water 188
Linnaea borealis 177
lixsgadim gayda ts'uuts' (EG) 19
lixsgedim gayda ts'uuts' (WG) 19
Lobaria pulmonaria 22, 159, 163
Lobaria spp. 22, 159
locoweed, mountain 177, 196
Lonicera involucrata 68, 155, 172, 174
Lontra canadensis 152, 159
loots' 71
Lotus denticulatus 184, 196

lung lichen 163
lupine 34, 148
lupine, arctic 98, 150, 177
Lupinus arcticus 98, 150, 177
luu gwiilahl ganaa'w 22
luux 60
Lycopodiaceae 29, 190
Lycopodium complanatum 29, 159, 183
Lycopodium dendroideum 31, 159, 182, 186, 190
Lycopodium obscurum 31, 182, 186
Lycopodium spp. 163
lynx 53
Lysichiton americanum 41, 188
Lysichiton kamtschatcense 145, 188
maa'y 98, 165
maa'yts 'Wii Gat (EG) 133
maa'yts 'Wii Get (WG) 133
maa'ytxwhl ganaa'w 124
maa'ytxwhl smax (EG) 55
maa'ytxwhl smex (WG) 55
maawin 27
maawint 27
madder family 200
majagalee 103, 196, 199
malgwasxw (EG) 146
Malte, M.O. 4, 17
Malus fusca 115, 150, 153, 156, 173
maple 27, 47, 61, 132
maple family 47
maple, Douglas 10, 47, 85, 153, 159, 175
meadowrue, western 109, 156, 159, 163, 175
medicine 4, 25, 32, 36, 39, 41, 49, 50, 61, 63, 71,
  75, 79, 84, 87, 101, 109, 115, 127, 132, 136,
  144, 146, 162
meega giist 63
melgwasxw (WG) 146
Mentha arvensis 103, 161, 163, 178, 185, 186
Mentha canadensis 103, 178, 185, 186
Menyanthaceae 197
Menyanthes trifoliata 177, 197
Menziesia ferruginea 91, 161, 186
Merrill, G.K. 17
Mertensia paniculata 177, 193
Microsteris gracilis 178, 198
microsteris, pink 178, 198
mii'isxwit 102
mii'oot 91
miigan 71, 93
miiganaa'w 124
miigwint 113
miik'ooxst 125
milk-vetch, alpine 185, 196
milkst 115
mint family 103, 197
mint, field 103, 161, 163, 178, 185, 186
miscellaneous 47, 49, 50, 66, 79, 109, 119, 124
miyahl (EG) 96
miyehl (WG) 96
monocotyledons 138, 149
Monotropa uniflora 171, 195
Moody, Joshua 14
moss 183
moss, peat 23, 153

mosses 22, 25, 152
mountain goat 21, 22
mountain-ash, Sitka 10, 127, 153, 156, 174
mountain-ash, western 10, 127, 153, 156
mountain-avens, yellow 113, 160, 183
mugwort, Michaux's 59, 161, 183
mushroom 19, 190
mussel 106
mustard family 68, 194
mustard, ball 185, 195
mustard, hedge 68, 155, 176
mustard, wormseed 179, 194
mythology 106
naasik' 121
National Herbarium of Canada 17
National Museum of Canada iii, 4, 11, 14
Neslia paniculata 185, 195
nettle 136
nettle family 136
nettle, stinging 10
Nicotiana tabacum 136, 189
nightshade family 136
nisk'o'o 124
North Wakashan 163
Nuphar polysepalum 104, 156, 162, 188
Nuxalk iii, 4, 11, 16, 43, 50, 63, 129, 136, 146
Nymphaea polysepala 104, 188
Nymphaeaceae 104
oleaster family 84
Onagraceae 106
onion 140
onion, nodding 10, 140, 151
onion, wild 188
Oplopanax horridus 55, 155, 188
orchid family 148, 202
orchid, one-leaved rein 148, 161, 184
Orchidaceae 148, 202
orpine family 82
Osmorhiza chilensis 53, 159, 163, 180
Osmorrhiza divaricata 53, 180
otter 109
otter, river 152, 159, 163
Oxycoccus oxycoccus 91, 150, 165, 173
Oxytropis monticola 177, 196
p'iinst 50
Pachistima myrsinites 77, 159, 163, 183, 186
Pachystima myrsinites 77, 183, 186
paintbrush, common red 133, 156, 181
Parmeliaceae 20
Parnassia montanensis 180
Parnassia palustris 180, 185, 197
Parnassiaceae 197
parsley family 49
parsnip, cow- 159
parsnip, wild 49
pearly everlasting 59, 157, 179
peavine, creamy 98, 150, 175
phlox family 198
Picea sitchensis 39, 188
Picea x lutzii 39, 150, 153, 155, 188
Pinaceae 36, 191
pine 25
pine family 36, 191

pine, jack 129
pine, lodgepole 10, 41, 150, 153, 155, 159, 188
pine, prince's 186, 195
pine, scrub 22, 39, 41, 132, 146, 188
Pinus contorta 41, 150, 153, 155, 159, 188
pixie cups, red 190
Platanthera obtusata 146, 148, 161, 184
Poaceae 148
Polemoniaceae 198
Polemonium pulcherrimum 177, 198
Polygonaceae 106, 198
Polygonum convolvulus 181, 198
Polypodiaceae 191
Polypodium glycyrrhiza 184, 191
Polypodium vulgare 184
polypody 191
Polyporaceae 19, 20, 190
pond-lily, yellow 10, 104, 156, 162, 188
poplar, balsam 129, 151, 154, 189
Populus balsamifera 129, 151, 154, 164, 189
Populus tremuloides 132, 151, 154, 156, 164, 189
Populus trichocarpa 129, 189
pore fungus family 19
potato family 136
Potentilla gracilis 179, 199
Potentilla viridescens 179
primrose family 199
Primulaceae 199
Prunella vulgaris 103, 161, 180
Prunus demissa 119, 171, 181
Prunus emarginata 117, 150, 172, 181
Prunus pensylvanica 117, 150
Prunus virginiana 119, 150, 153, 160, 171, 181
Pseudotsuga menziesii 188, 191
Pseudotsuga mucronata 188
Pteridium aquilinum 25, 155
pteridophytes 25, 149, 190
puffball 19
pussytoes, field 182
pussytoes, rosy 176
Pyrola asarifolia 184, 195
Pyrus diversifolia 115, 173
Pyrus occidentalis 127, 174
Pyrus sitchensis 127, 174
Queen's cup 140, 158, 162, 181
rabbit 140
Ramaria 19
Ranunculaceae 107, 199
Ranunculus abortivus 108, 109, 152, 159, 163, 176
Ranunculus repens 186, 199
raspberry, five-leaved creeping 124, 151, 160
raspberry, red 10, 102, 121, 151, 178
raspberry, trailing 10, 124, 151, 160, 163, 175
rattle yellow 181, 201
raven 162
redcedar 34, 36, 39, 43, 61, 75, 99
redcedar, western 10, 34, 153, 157, 164, 188
Rhinanthus crista-galli 181
Rhinanthus minor 181, 201

Ribes bracteosum 101, 188
Ribes hudsonianum 102, 149, 150, 165, 175, 180
Ribes lacustre 101, 102, 156
Ribes laxiflorum 101
Ribes oxyacanthoides 102, 150, 185
rice-root, northern 10, 141, 151, 180
riceroot 142
ritual/spiritual 49
Robinson, Bob 13, 19, 23, 31, 49, 50, 53, 59, 60, 68, 77, 84, 89, 91, 101, 102, 103, 106, 107, 108, 109, 112, 113, 129, 141, 144, 148, 190, 191, 192, 193, 195, 196, 197, 198, 199, 200, 202
rockcress, hairy 176, 182, 194
Rosa acicularis 121, 151, 153
Rosa nutkana 121, 151, 153
Rosa sp. 121, 173
Rosaceae 109, 162, 199
rose 10, 27, 75, 121
rose family 109, 199
rose, Nootka 121, 151, 153
rose, prickly 121, 151, 153
rose, wild 173
Rubiaceae 200
Rubus idaeus 121, 151, 178
Rubus parviflorus 124, 151, 160, 173
Rubus pedatus 124, 149, 151, 160
Rubus pubescens 113, 124, 149, 151, 160, 163, 165, 175
Rubus spectabilis 125, 151, 165, 178
Rumex acetosella 106, 150, 179
rush, branched horsetail 29
rush, horsetail 27
rush, scouring- 29, 161, 175
Salicaceae 129
Salix sitchensis 132, 160, 176
Salix sp. 132, 154, 176
salmon 34, 39, 43, 61, 66, 71, 80, 91, 96, 98, 99, 103, 107, 110, 112, 113, 122, 124, 129, 133, 140, 141, 144
salmonberry 10, 125, 151, 178
Sambucus racemosa 71, 93, 150, 155, 174
Sampare, Gus 14, 29, 59, 82, 101, 190
Sampare, Mrs. Gus 14
Sampare, Mrs. Robert A. 14, 199
Sampare, Robert A. 14, 22, 82, 106, 141, 199
sandalwood family 133
Santalaceae 133
sarsaparilla, wild 55, 152, 159, 163, 176
Saskatoon 10, 34, 43, 110, 150, 153, 172, 187
Saxifraga tricuspidata 177, 200
Saxifragaceae 200
saxifrage family 200
saxifrage, three-toothed 177, 200
saxsduu'lxw 141
Scrophulariaceae 133, 201
sdatxs (EG) 136
sdetxs (WG) 136
Sedum ?stenopetalum 82
Sedum divergens 163
Sedum lanceolatum 82, 155, 163
Sedum stenopetalum 82, 177, 181

seeks 39
Sekani 4
self-heal 103, 161, 180
Senecio balsamitae 177
Senecio cymbalarioides 179, 192
Senecio pauperculus 177, 193
Senecio triangularis 184, 193
sgan'isxwit 103
sgandaxdo'ohl 89
sgandilusa'a 102
sgangam (EG) 110
sgangapk'oyp 141
sgangem (WG) 110
sgangisgits 73
sgangisgits' 73
sganhlingit 127
sganis 84
sgank'alamst 121
sgankw'ats 60
sganloots' 71
sganmaa'ya smax (EG) 107
sganmaa'ya smex (WG) 107
sganmajagalee 197, 200
sganmiigan 93
sganmiiganaa'w 124
sganmiigwint 113
sganmilkst 115
sganmiyuxws 59
sgannaasik' 121
sgannaxnok 32
sgannisk'o'o 124
sgansaxsduu'lxw 141
sgansnaw 117
sgansnax 112
sgant'imi'yt 87
sgants'ak' (EG) 66
sgants'ek' (WG) 66
sgants'idipxst 75
sgants'iks 146
sgantya'ytxw 91
sganxhlaahl 79
sginist 41
sheep 106
shepherd's purse 182, 194
Shepherdia canadensis 84, 150, 155, 173
Sisymbrium incisum 176
Sisymbrium officinale 68, 155, 176
skunk-cabbage 10, 146, 188
Smilacina racemosa 17, 144, 151, 156, 172, 175, 180
Smilacina sp. 141, 180
Smilacina stellata 145, 161, 178
snaw 117
snax 112
snowberry, common 10, 73, 152, 153, 171
soapberry 10
Solanaceae 136
Solidago canadensis 187, 193
Solidago lepida 187
Solidago sp. 174
Solomon's-seal, false 17, 144, 151, 156, 172, 175, 180

Solomon's-seal, star-flowered false 145, 161, 178
Sonchus arvensis 60, 161, 163, 184
soopolallie 39, 47, 84, 124, 150, 155, 173
Sorbus scopulina 127, 153, 156
Sorbus sitchensis 127, 153, 156, 174
sorrel, sheep 106, 150, 179
sow-thistle 163
Sphagnidae 22
Sphagnum angustifolium 22
Sphagnum capillaceum 22
Sphagnum fuscum 22
Sphagnum girgensohnii 22
Sphagnum sp. 23, 185
Sphagnum spp. 22, 153
Sphagnum squarrosum 22
sphagnum, common brown 22
sphagnum, common green 22
sphagnum, common red 22
sphagnum, poor-fen 22
sphagnum, shaggy 22
Spiraea douglasii 129, 161, 181, 183, 184
spruce 10, 25, 34, 39, 41, 84, 146
spruce, hybrid Sitka 10, 39, 150, 153, 155, 188
spruce, Sitka 10, 39, 188
squashberry 50, 75, 121
stafftree family 77
starflower, northern 178, 199
Stictaceae 22
stinging nettle 136, 154, 156, 174
stonecrop, lance-leaved 82, 155
stonecrop, spreading 163
stonecrop, worm-leaved 82, 177, 181
strawberry 113, 121
strawberry, wild 10, 113, 150
Streptopus 141
Streptopus amplexifolius 145, 161, 175
sweet-cicely 112
sweet-cicely, mountain 53, 159, 163, 180
Symphoricarpos albus 73, 152, 153, 171
Symphoricarpos racemosus 73
Symphoricarpus racemosus 171
Sysymbrium incisum 68
t'imi'yt 87
t'ipyeeshl gaak 82
Tamias amoenus 152
Taraxacum ceratophorum 193
Taraxacum officinale 193
Taxaceae 45
Taxus brevifolia 45, 153, 188
tea, Labrador 10, 89, 150, 155, 162, 185
Thalictrum occidentale 49, 50, 109, 156, 159, 163, 175
thimbleberry 10, 84, 101, 107, 113, 122, 124, 133, 151, 160, 173
thistle, sow- 60, 161, 184
thistle, wavy-leaved 183, 192
Thuja plicata 34, 153, 157, 164, 188
Tiarella trifoliata 186, 200
Tiarella unifoliata 186
tl'ok'ats 107
toad 163

toad-flax, bastard 133, 158, 162, 171
tobacco 87, 136, 189
tooth fungus family 18
trefoil, meadow bird's-foot 184, 196
<u>Trientalis arctica</u> 178, 199
trout 112
ts'anksa gaak (EG) 140
ts'enksa gaak (WG) 140
ts'idipxst 75
Tsimshian 5, 11
Tsimshian, Southern (Kitasoo) iii
Tsimshianic iii, 163, 165
<u>Tsuga heterophylla</u> 43, 153, 164, 188
turtlehead 187, 201
twinberry, black 68, 155, 172, 174
twinflower 177
twistedstalk, clasping 145, 161, 175
<u>Ursus americanus</u> 152, 159
<u>Ursus arctos</u> 152
<u>Ursus</u> spp. 152
<u>Urtica dioica</u> 136, 154, 156, 174
<u>Urtica Lyallii</u> 136, 174
Urticaceae 136
<u>Vaccinium alaskaense</u> 93, 150
<u>Vaccinium caespitosum</u> 96, 150, 165, 176, 179, 186
<u>Vaccinium membranaceum</u> 98, 150, 165
<u>Vaccinium ovalifolium</u> 93, 150
<u>Vaccinium oxycoccus</u> 91, 173
valerian family 138
valerian, marsh 138, 154, 178
<u>Valeriana dioica</u> 138, 154, 178
<u>Valeriana septentrionalis</u> 138, 178
Valerianaceae 138
<u>Veratrum viride</u> 145, 156, 162
<u>Viburnum edule</u> 75, 150, 153, 155, 171
<u>Viburnum pauciflorum</u> 75
Victoria Memorial Museum 20
<u>Viola adunca</u> 176, 201
<u>Viola canadensis</u> 176, 201
Violaceae 201
violet family 201
violet, Canada 176, 201
violet, early blue 176, 201
waasan 132
waasanhl ts'imilix 132
Washburn, W.C. 14
water-lily family 104
wheatgrass, slender 148, 154, 178
Wiget 133, 140
wildrye, blue 148, 154, 178
willow 10, 22, 34, 129, 132, 133, 154, 164
willow family 129
willow, beaver 133
willow, Sitka 132, 160, 176
wintergreen, pink 184, 195
woodchuck 39
xsduu'lixs 145
xwdakw 45
yans (EG) 193
yarrow 10, 57, 155, 174
yens (WG) 193

yensa giist 63
yew 45, 188
yew family 45
yew, Pacific 45, 153, 188
yew, western 10, 45, 153, 188